The Hazards of Life and All That

# The Hazards of Life and All That

A Look at some Accidents and Safety Curiosities,

Past and Present

John Bond

with Illustrations by Keith D Jenkins

Institute of Physics Publishing
Bristol and Philadelphia

*British Library Cataloguing-in-Publication Data*

A catalogue record for this book is available from the British Library.

ISBN 0 7503 0360 3

*Library of Congress Cataloging-in-Publication Data*

Bond, John.
The hazards of life and all that : a look at some accidents and safety curiosities, past and present / John Bond ; with illustrations by Keith D. Jenkins.
p. cm.
Includes bibliographical references and index.
ISBN 0-7503-0360-3
1. Industrial safety. 2. Industrial accidents. I. Title.
T55.B59 1996
363. 11-- dc20 96-28766
CIP



Published by Institute of Physics Publishing, wholly owned by The Institute of Physics, London

Institute of Physics Publishing, Techno House, Redcliffe Way, Bristol BS1 6NX, UK

US Editorial Office: Institute of Physics Publishing, The Public Ledger Building, Suite 1035, 150 South Independence Mall West, Philadelphia, PA 19106, USA

Typeset in the UK by Mackreth Media Services, Hemel Hempstead, Herts
Printed in the UK by J W Arrowsmith Ltd, Bristol

# ACKNOWLEDGMENTS

The author wishes to thank all those who have known of his interest in the subject and have been generous in providing stories. He is indebted to his sister, Margaret Bond, for some anarchic comment and historical gleaning. He is especially grateful to Patricia, his ever-tolerant wife, for her immense help and encouragement in bringing this book together. To her is dedicated that hazard of matrimonial bliss recorded by Samuel Pepys on New Year's Day 1662:-

> 'Waking this morning out of my sleep on a sudden, I did with my elbow hit my wife a great blow over her face and nose, which waked her with pain — at which I was sorry. And to sleep again.'

# CONTENTS

# FOREWORD

It is indeed an honour to write this foreword to a book authored by my colleague Dr. John Bond.

I met John many years ago in London and before that enjoyable first meeting was over, I knew this man was truly the safety historian of the world.

During a long career, it has been my great privilege to travel extensively as a consultant and teacher of industrial safety. While I have met hundreds of fine safety leaders around the world and have also been privileged to know a significant number of the pioneering giants in our great profession, Dr. John Bond holds a very special niche in my memory bank of great contributors to our profession.

He has long demonstrated an unusual in-depth knowledge of the causes and control of accidents through his many outstanding written and oral contributions to internationally respected organisations and journals.

This book is another of John's many contributions that truly represent the depth of his passion for safety. This anthology of safety and accident epigrams and other unique sayings is timeless.

Just as John would hope, I am confident this contribution will serve his intended purpose to enhance the communications of thousands upon thousands of professionals around the world in their loss prevention efforts.

It is no exaggeration to predict that this interesting and unequalled historical collection of lessons from loss will become a treasured part of every safety professional's library of useful references.

**Frank E. Bird, Jr**
Founder and former President
International Loss Control Institute

# INTRODUCTION

MILLIONS OF CHILDREN in the English-speaking parts of the world have grown up with an awareness of the hazardous nature of life, and a foretaste may well have drifted into the subconscious with the lullaby:-

*Hush-a-bye baby on the tree top;*
*When the wind blows the cradle will rock;*
*When the bough breaks the cradle will fall,*
*Down will come baby, cradle and all.*

A little later toddlers learn that:-

*Jack and Jill went up the hill*
*To fetch a pail of water;*
*Jack fell down and broke his crown,*
*And Jill came tumbling after.*

And when able to climb they are warned:-

*Humpty Dumpty sat on a wall,*
*Humpty Dumpty had a great fall.*
*All the King's horses and all the King's men*
*Couldn't put Humpty together again.*

Children then learn to play with each other and commemorate — whether or not they know it — a calamity of the past, the Great Plague. They chant joyfully:-

*Ring-a-ring of roses,*
*A pocket full of posies.*
*Atish-shoo, atish-shoo,*
*We all fall down.*

They also learn about an important occurrence in Medieval London:-

*London Bridge is falling down, falling down, falling down;*
*London Bridge is falling down, my fair lady.*

But it has always struck me that our nursery rhymes are very negative in their approach to safety matters, and perhaps children ought rather to be brought up with a more constructive view of the pitfalls they will encounter in life. Admittedly it would take a lyricist of remarkable genius to do anything with the sentiments expressed in the following:-

*Hush-a-bye baby on the tree top;*
*When the wind blows the cradle will rock;*
*But since it has been secured to the International Standard,*
*The probability of it falling is greatly reduced,*
*Although an inherently safer system would be at ground level*†.

There are, however, other ways to ensure that children grow up with the right approach to safety. I was lucky. Once I was old enough to understand the matter, my mother warned me about the hazards of fire. She did not take my hand and plunge it into the flames on the theory that a burnt child avoids the fire. Instead she explained that the fire would hurt if I allowed it to touch me, but she also put a guard in front of the fireplace so that I could not even reach it. Maternal instinct coincided with today's belt-and-braces philosophy, and two safeguards were provided against an identified hazard. Moreover, these safety precautions were closely monitored; in other words, she kept an eye on me.

This book records some of the more unusual accidents and safety oddities that have come my way. Many of those from the past were not, of course, considered unusual at the time, however bizarre and amusing they may seem today. Nevertheless they can serve as a reminder that every age has its own stock of hazards waiting to be overcome. It is perhaps going too far to claim that a lesson can be drawn from each of the mishaps I have included, but it is just possible that by knowing more about accidents the reader will avoid becoming the victim of one. Pessimists will take lugubrious pleasure in pointing out that even though the Middle Ages recognised the danger of stowing rubbish where it might present a fire hazard the practice has never stopped. But this is no reason why the point should not be hammered time and time again.

ACCIDENTS LEAD TO LOSS — loss of life, loss of property, loss of earning power by individuals and corporate bodies. We who do our soldiering in the Loss Prevention Brigade have two laws. The First Law states:-

†A Factory Inspector of the Health and Safety Executive suggests:-

*Hush-a-bye baby on the tree top:*
*When the wind blows the cradle will rock;*
*Approved Standard fittings should thwart a great fall,*
*But a ground level manger is safest of all.*

'He who ignores the past is condemned to repeat it.'

but because studying what *has* happened is rarely sufficient we also consider what *could* happen, and so the Second Law states:-

'He who anticipates the future can safeguard the present.'

I speak for many when I take this opportunity of paying tribute to such bodies as the Health and Safety Executive, the International Maritime Organisation, the Institution of Chemical Engineers, The Royal Society of Chemistry, the Royal Society for the Prevention of Accidents, the British Safety Council, the International Loss Control Institute and many others for their tireless efforts to make life safer for all of us. Publicity is always given to the accidents that actually take place — never to the ones that have been prevented by the vigilance of such organisations.

A FINAL WORD. A book of this nature is no place for the many tragic accidents of recent times and I have made no mention of them.

**John Bond**

# CHAPTER 1

# HORS D'OEUVRE

**ACCIDENTS ARE OF TWO KINDS:
THOSE OCCASIONED BY ACTS OF GOD AND THOSE CAUSED BY MAN.**

ACTS OF GOD include such events as thunder storms, earthquakes, volcanic eruptions, hurricanes and tidal waves. We cannot prevent them, but we can and do (or should) take steps to mitigate their effects — lightning conductors and gale warnings to shipping, for instance, have done much to prevent loss of life from hazards over which we can have no control. The accidents caused by man are usually the result of negligence or ignorance, although malevolence plays its part in many past and present misadventures. Even accidents arising from man's encounters with the animal world can be attributed to his failure to take precautions against the Nature of the Beast.

What is perhaps most striking in the long list of recorded accidents and mishaps is their sheer variety, some predictable from the dangerous nature of the circumstances, others the result of seemingly innocent activity. What, for instance, are we to make of this sad little epitaph from New Jersey in the United States:

*Julie Adams*
*Died through wearing thin shoes*
*April 17th 1839*
*Aged 19 years*

A victim of a repeated Act of God was Major Walter Summerford, whose experience with lightning must surely defy the laws of probability. It is reported that he was struck by lightning on the battlefield of Flanders in 1918. As a result of the injuries he sustained he was invalided out of the army. In 1924 while in Canada he was again struck by lightning, and yet again in 1930, when he became paralysed. He died in 1932 and was buried near Vancouver. His gravestone was later struck by lightning and destroyed.

What was described as 'a most extraordinary and awful circumstance' occurred in the village of Saint Chamand in France in June 1836. The wife of the local doctor had died and the relations had assembled to pay their last respects to the deceased and attend the funeral.

When the body was being placed in the coffin the undertakers noticed some slight signs of life; the shroud was removed and the woman began to revive. At that moment a violent storm broke and a bolt of lightning struck the house, killing the doctor's sister and a servant. All attempts to maintain the revival of the doctor's wife were fruitless and 'her funeral was adjourned to the following day; instead of one corpse, three were then conveyed to their last home.'

A SCHOOLBOY HAD A LESS MACABRE ENCOUNTER with lightning. Writing to his parents from Eton in the early 1820s, John Mordaunt reported:

'I believe I ought to have written before to tell you of the narrow escape which I had yesterday, for which I am very thankful. I daresay you will like to know the particulars as there is no cause for alarm. It was from a thunderbolt which fell on this house about half past 2 o'clock; it came in at a small passage, (from which I was about 2 yards) ran along the wire of a bell in the contrary direction from me, and passed out again in the room beyond, having knocked down a great part of the wall; from the immense shock at the time I was knocked across the room and rather stunned, as if it had gone through my head; but I soon recovered, and knew directly what it was; you must conceive my feelings better than I can describe them. When I looked up, the room was filled with smoke, and I thought that the house was coming down, or on fire, in short I hardly knew what. Luckily the house did not catch fire, but the part which it entered, was very much knocked down, and covered with black, everybody else was in the room below at dinner from which I had just come up; And if it had come a moment sooner it might have hit me as I was coming up the stairs. It was a fire ball, but I do not know whether this is the same as a thunderbolt.'

Wider publicity was given to a thunder storm in 1680, when a pamphlet was 'Published for prevention of false Reports'. It guaranteed 'A full and true Relation of the Death and Slaughter of a Man and his Son at Plough, together with four Horses, in the Parish of Cookham in the County of Berks, Sept. 2, 1680, slain by the Thunder and Lightning that then and there happened, as may be fully be testified by credible Persons, whose Names are hereunto adjoyned.' The pamphlet began:-

> 'Thursday, September the Second, 1680. John Sayer, a Farmer with his Son Richard, a Youth of 13 Years of Age, went with four Horses halfe a mile to a Field in the Parish of Cookham, called by the Name of Ham Field, to Plough, about the hour of Ten in the forenoon, it being then fair weather, but about Eleven a Clock the Sky began to lower, the Clouds grew thick, and soon it Lightned and Thundred, and some showers of Rain fell, it thundred several times very loud, sharp and shrill, to the amazement of several Persons: two other Ploughs were at work in the same Field at the same time, their Cattel being afrighted and unruly, they shoot off and go home, and leave John Sawyer in the Field, about one of the Clock the Tempest began to give off.'

Four hours later a passing labourer arrived on the scene and:-

> 'Sees John Sawyer, with his Son Richard, and the four Horses all dead on the ground: knowing the Man and his Cattel, he came into Cookham, and acquaints the Inhabitants with the Matter, immediately the Major part of the Towns Folks flockt up into the Fields to see the sad Spectacle of amazement, and with the rest, the Wife of the said John Sawyer, where to her exceeding grief she saw her Husband, Son, and four Horses dead, no impression of any stroak or bruise on the Man, but the Boy his Cloaths were most part of them rent from his body, and his Hat torn into two or three Pieces; One Shooe all the upper leather torn from the Sole and Heel: and his Whip broke in two or three pieces, his Shirt beaten to Lint, and strewed on the Horse he was supposed to ride, except one narrow Slip from his Neck to the fore part, only remaining from his Neck downward before; he was observed to be naked, and from the Neck down his Back burnt, or singed, as is supposed with the Lightning, likewise scorched or singed on the belly, some part of his hair singed, as the Eye-brows, and part of the hair of his Head; and that the Horse he was supposed to ride, was singed from the Neck to the Flanck.'

The report ended by stating that:-

> 'John Sawyer was a Man of good repute among his Neighbours, that lived Soberly and Honestly. On Saturday the 14th of September, the Coroner came and called a Jury for Inquiry, and they gave their Verdict that it was the immediate providence of

Almighty God; and so the Coroner gave order for the Burial of the Man and his Son, which the same Evening was performed.'

IN THE TOWN OF LEWES, Sussex, is a pub with the charming name of Snowdrop Inn. The name has nothing to do with the lovely little flower, but marks the site of the worst avalanche in British history. A hill had been excavated for chalk, leaving a sharp drop of some one hundred metres to where a row of houses stood. On Christmas Eve of 1836 there was an unusually heavy fall of snow, with strong winds. By Christmas Day it had stopped snowing but was freezing hard, and the wind-driven snow on the hill was now a great frozen mass poised over the houses below. On Boxing Day a small part of it came down on a nearby timber yard, but the residents even then could not be persuaded to leave their homes. On the 27th December the rest of it fell. Eight people were killed and are buried in the nearby church, where a plaque records 'the awful instance of the uncertainty of human life.'

THE AUTHOR'S CURIOSITY was recently aroused by a carefully painted wooden sign he came across in an antique warehouse:-

*Do not spit on the floor.*
*Remember the Jamestown flood.*

The relevance of this sign may seem obscure, but the Jamestown flood was probably a misquote for the Johnstown flood which has a special place in the folk memory of the people of Pennsylvania. The flood occurred in 1889, when the South Fork reservoir, 12 miles above the booming steel town of Johnstown, broke through an earth dam 1,000 ft long and 100 ft high. The dam and artificial lake were privately owned as part of a fishing and hunting club, and the spillways had been blocked to prevent the loss of fish! The water level had been rising for two days when, on the 31st May, the clouds burst and three inches of rain fell in 20 minutes. At 3 p.m. the dam gave way and 4½ billion gallons of water were unleashed on Johnstown, 400 ft below the dam and 12 miles distant. There had been so many doomladen but unrequited predictions that the dam was about to give way that when it actually did so the residents ignored a frantic citizen who galloped down with the news, urging them to take to the hills. The death toll was variously estimated at between 2,500 and 7,000. The bar room sign and its tasteless reference to the tragedy was evidently intended as a waggish reminder that instructions must be heeded — or else.

FALLING ROCKS CAN BE A DANGER in many mountainous areas and geologists know the type of structure which is particularly susceptible to fracture. This knowledge was not available to a shepherd, who, in 1760, took his bride to his cottage at Gribun on the Isle of Mull after celebrating their marriage at a local farm. During the night a large piece of rock detached itself from the mountainside and crushed the cottage. The couple are buried under the stone which is still partly surrounded by the garden walls of the cottage.

Another rockfall, also with fatal consequences, was the work of vandals. As reported in August 1797:-

> 'Some miners from the tunnel of the Huddersfield canal went for amusement, with a great number of people, to blow up a very large rock in Greenfield, in Saddleworth, known by the name of Raven Stone. After many fruitless attempts, they effected their purpose in the evening, when they tore that venerable relic of antiquity from its ancient basis. It fell with a most dreadful crash and, dividing, took different directions. We are sorry to add, that one man was killed upon the spot, and others so much wounded that they are not expected to recover. The rock had been long admired for its towering grandeur, and had near a mile to roll down a very steep eminence.'

The ancient megalithic rocks at Avebury in Wiltshire were buried underground in the 13th century, probably because local superstitions associated with them were frowned on by the Church. Holes were dug adjacent to the vertical stones which were then pushed into the pits and covered with earth. When the site was excavated in 1938 the remains of a man were found beneath a 35 ton megalith. It is assumed that the stone fell onto the unfortunate man who was digging the hole. He may have been a local tailor pressed into service, as a pair of scissors were found with him.

But a more common hazard facing tailors has always been the swallowing of pins foolishly held between the teeth, and those who sew and stitch can take warning from a memorial tablet in the parish church at Dedham in Essex. It reads:-

> 'In memory of Judith (Coyte) Eyre,
> wife of Joseph Eyre, gentleman,
> many years of this Parish, who died
> much lamented in the 35th year of
> her age, January 25th 1747/8, in

consequence of having accidently
Swallowed a pin.'

Negligence accompanied by ill will caused the death of an unfortunate woman in 1326. The Coroners Rolls record that:

'On Monday, the Feast of the Nativity, John Rynet and Alice his wife were alarmed at midnight by a fire which had been caused by the fall of a lighted candle, as they were going to sleep, and hurriedly left the burning shop. Immediately afterward John, blaming Alice for causing the disaster, violently pushed her back into the shop and fled, but whither the jurors knew not. Alice was thus injured by the fire and again leaving the shop, lingered until the following Tuesday, when she had her ecclesiastical rights and died of burns.'

A SINGULAR ERUPTION early this century left the residents of Boston, Massachusetts, with a sticky situation on their hands and is a reminder that even the most agreeable of substances can be lethal if not safely contained. The substance was treacle and it was contained in a 50 ft high tank of dubious construction. On 15th January 1919 the tank burst and 12,500 tons of treacle poured out like lava from a volcano. Men, women and children were engulfed, timber houses and trucks were swept away, horses were drowned. Twenty-one people were killed and some forty injured. Bodies were still being recovered a week later.

Anybody who has ever spilled a jar of treacle, or molasses as it is known in the United States, will appreciate the appalling mess. Rescuers and nurses on the scene became coated with the syrup; the whole city was soon contaminated via the sticky public transport and hardly a telephone booth in Boston was unsullied. Hoses using the public water supply were ineffective in swilling away the ocean of treacle, although sea water used by the fire service proved more helpful. But how do you pump out a cellar full of treacle?

**ACCIDENTS TO HUGE TANKS** did not always have such unwelcome consequences. On 14th October 1814 one of the vats at a brewery in London burst and deluged surrounding streets with its contents 'amounting to 4,500 barrels of strong beer.' The brewery walls collapsed and three of the employees 'were rescued with great difficulty by the people collected to afford relief, who had to wade up to the middle through the beer'. Two adjoining houses were demolished and because the site was low-lying nearby cellars were flooded. The press report continued:-

> 'When the beer began to flow, the neighbourhood, consisting of the lower classes of the Irish, were busily employed in putting in their claim to a share, and every vessel, from a kettle to a flask, were put into requisition, and many of them were seen on Tuesday enjoying themselves at the expense of the proprietors, whose loss is estimated at an immense sum.'

Even the great and the good need to be wary of vats. The economist Adam Smith, while Professor of Moral Philosophy in Glasgow, visited a tannery with a friend. Standing on a plank laid over a tanning pit and talking excitedly, he stepped to one side and fell head first into a dye-bath. He was pulled out, stripped, wrapped in blankets and carried home in a sedan chair, complaining that he was dying from the cold.

**DELIBERATELY SETTING FIRE** to schools is not, as we may be inclined to think, a modern habit of disaffected youngsters. The *New Annual Register* for 1789 reported that on July 10:

> 'Mr. Williams and family of Bratton school, were alarmed from sleep by a dreadful fire, which burst out from the roof of their house. The consternation occasioned by this event, at such a time, is inexpressible. By the mercy of Providence, however, not a single person of nearly fifty was hurt. Mr. Williams has sustained

a heavy loss, but two neighbouring gentlemen have afforded him substantial assistance; one, by accommodating him with a large house at Westbury, for the immediate reception of his scholars; the other, by beginning a subscription with a liberal donation.'

But a sequel was reported on September 17:

'Mr. Gibbs of Westbury, who generously accommodated Mr. Williams of Bratton-school, with the use of his house has since had it burnt down. This created a suspicion that one of the boys must be the incendiary, which has proved the fact. The boy on whom the suspicion fell has made a confession of his having accidentally set fire to the school-house at Bratton, and wilfully to that of Mr. Gibbs. On his examination he said, 'the thought unluckily came into his head, that, if he could burn the school-room at Westbury, he might be sent home, to which his father had not permitted him to return for fifteen or sixteen months past.' He was committed to Devizes. [A few days after, he put an end to his existence.]'

AKIN TO THE BLUNDER which led to the disastrous charge of the Light Brigade at Balaclava in 1854 was an error that made it necessary to abort a space probe. The Mariner 1 probe to Venus was blasted off from Cape Canaveral in 1962 but shortly afterwards it veered off course and had to be destroyed before diving into the Atlantic shipping lanes. The subsequent investigation revealed that part of a computer program had been incorrectly written: a minus sign had been omitted.

GREEN ARCHITECTS AND BUILDERS are nowadays growing enthusiastic over the use of grass turf as a roofing material. There is nothing new in the idea, as house roofs in the Netherlands have been covered in this way for many years, and it is from that country that a pointer to the danger comes. In October 1695 Lord and Lady Kilsyth visited Amsterdam. They brought disaster with them, and shortly after their arrival parts of the house they were staying in collapsed, killing Lady Kilsyth, her son and a chambermaid. A member of the party explained in a letter what had happened:-

'... the people to whom the house belonged had bin, all that day, carreing up theer turfs to the chamber immediately above thers, &, after they had carried up the last sackfull of 300 [sic] tuns, the weight of that quantity of turff broke doune the loft above them...'

RECENT FIRES HAVE SHOWN the danger of stowing rubbish away out of sight. This foolish practice probably started when man stopped being a nomad, and in a frugal community there may be some excuse.

A fire broke out in 1198 at the monastery of Bury St. Edmunds. It is related that there was a wooden platform, on which lighted candles stood, between the High Altar and a box containing sacred relics. 'Under this platform many things were stored without regard to seemliness, such as flax, thread, wax, and divers utensils; in fact, anything that came into the hands of the guardians was placed there, since the platform had a door and iron walls.' On the night of the Feast of St. Ethelreda a candle burned down, fell on the cloth covering the platform, and 'began to set fire to everything near it both above and below, so that the iron walls were white with the heat... And we all of us ran together, and found the flames raging beyond belief and ... reaching up nearly to the beams of the Church...'

An improbable item of rubbish tidied away almost led to a serious fire at the House of Commons in 1792 and raised suspicions of sinister intent. Viscount Palmerston (father of the Prime Minister) wrote to his wife on the 10th May:

> 'A strange incident happened last night at the House of Commons some hours after it was adjourned. Some clerks who were writing, smelt a violent smell of fire with smoke which they could not account for and after much search they at length found in a kind of cupboard in a water-closet below the House of Commons a pair of old fustian breeches burning and stinking at a most extraordinary rate... There was not much of them consumed but they were a considerable time burning and made a great stink and smoke. I am not inclined to look on this as a serious plot...'

Lady Palmerston in reply suggested that it might be a conspiracy of the *sans culotte* party.

Another House of Commons, this time of the Dublin Parliament, was the scene of a more serious fire and it is comforting to note that even in an emergency the correct Parliamentary procedure was followed. The *New Annual Register* of 28th February 1792 reported:

> 'At half past five yesterday evening, as the House of Commons were in a committee, they were panic-struck by a voice from one of the ventilators at the top, communicating the dreadful intelligence that the roof was in flames, and the dome would fall

> within five minutes. The Speaker instantly resumed the chair, and put the question of adjournment: the deliberative faculty of the house was immediately suspended: and every member escaped as he could with the utmost precipitation...
>
> 'Every necessary precaution was used by the Speaker to preserve the books and papers of the House...'

Most fires can be traced back to a single cause. A painstaking investigation into why a fire alarm had sounded in a drawing office when no fire was obvious came up with an unusual little culprit. It became apparent that a box of non-safety matches had ignited and set fire to some papers, setting off the alarm. Meticulous examination showed tiny teeth marks on a match lying on the floor under the desk. It was concluded that a mouse had got into the drawer and ignited a match, possibly by trying to separate it from another it was fused to.

A pet hamster playing with a match also caused a fire recently, this time at a cottage in the country. It set fire to the wood shavings in its cage and the blaze damaged the top floor of the cottage. The hamster, badly singed, survived the ordeal and because it enjoyed chewing on wood, the owners were advised by the fire authorities to give it only spent or safety matches to play with.

Nor can non-safety matches be blamed for all fires in which little furry animals are implicated. Some insurance experts now believe that one fifth of all domestic fires of unknown origin are caused by rodents gnawing electric cables; moreover, the chewing of plastic pipes in central heating systems quite commonly results in flooding.

And not only domestic fires — after a severe fire at an electrical engineering factory investigators found the burned bodies of a cat and a rat embedded in the switchgear of a transformer testing bench. They concluded that the cat had been chasing its natural prey and that both had come into contact with live terminals behind the bench.

Fires are not the only consequence of rodent activity. In 1840 a commission set up to consider child labour in coal mines found that children were sometimes employed to operate machinery, a responsibility they were far too young to undertake. On one occasion three people were killed when a child of nine, operating a hoist machine, turned away from his work to look at a mouse.

**A CORONER'S REPORT FROM 1321** pointed to the danger of not looking where you are going:-

'When on Sunday at dusk, Elena Scot, a servant, left the solar of the house to get some fire, she slipped from the top step of the entrance of the solar and fell backwards down the steps upon a stone at the bottom and broke her neck and forthwith died in consequence of that and from no other felony. Being asked who were present when this happened, the jurors say Margaret de Sandwich, her mistress, and one Christina Lovel, and Margaret first discovered the corpse and raised the cry, so that the country came; nor do they suspect any man or woman of the death, but only mischance. The corpse was viewed, on which the broken neck appeared and no other hurt.'

AS WILL BE SEEN from our chapter on accidents at work, carrying dangerous liquids can be a dicey business and the problems are compounded when carriage is by public transport. In 1759 it was reported that the Worcester Wagon was burnt out with the loss of £5,000 owing to the bursting of a bottle of aquafortis [nitric acid] placed among the baggage.

Nitric acid made its mark again in Paris in 1851. A bus was passing through the Place de la Bastille when volumes of smoke were seen issuing from it. The passengers were quick to get off. Inside the bus a man rolled about in great pain; hardly surprising, since he had been carrying a bottle of nitric acid in his pocket and a jolt of the bus had caused the bottle to break. Apart from his own burns, the clothes of a woman seated next to him were burnt and the matches in the pocket of another man caught fire.

*Plus ça change...* fifty years later public transport in Paris was still an unsafe means of travel. On 8th October 1901 six passengers on a Paris tramcar were suddenly seized with violent abdominal pains. The tramcar was stopped and the suffering passengers were treated at a nearby chemist shop. An inquiry was held and the opinion expressed that the poisoning was caused by the fumes given off when sulphuric acid attacked the lead in the accumulators.

The list of hazardous substances that it is inadvisable to carry on public transport grows daily. In October 1991 the main east coast railway station at York was closed for two hours after a refrigerated flask containing human sperm fell off a trolley, releasing liquid nitrogen. A detail that must chill the blood of any Welsh Nationalist notes that the sperm was being sent from Doncaster to South Wales for *in-vitro* fertilisation.

WE HAVE NOT included any motoring accidents in our chapter on the hazards of travel because they are too numerous — even commonplace — to contemplate here. But we cannot resist the tale of a collision not between two cars but between their drivers. It happened at Gutersloh in Germany a few years ago, when two drivers approaching each other in thick fog were craning their necks out of their windows in an attempt to see the white centre line. They banged their heads together and both needed hospital treatment.

At least the windows in that little incident had been wound down, but in a letter dated 21st February 1744 the accomplished Mrs. Delaney gossiped to her sister:-

> 'Yesterday as my Lord Winchelsea was going to Court the glass was up and his blind eyes did not perceive it, so that bowing to somebody, and the coach giving a great jolt at the same time, popped his head quite through, and has cut his forehead violently; it is well he did not lose an eye.'

**IT HAS BEEN WELL SAID** that prevention is better (and cheaper) than compensation. From a report dated 25th June 1790 we learn that:-

> 'This day an action was brought by Mr. Norris against Mr Goodman, a baker, in Coventry-street, for not keeping the iron plate which covers the mouth of the coal-hole in the street, in repair, whereby the plaintiff fell down and broke his thigh. It appeared that a small part of the iron plate was worn away, and that as the plaintiff was walking along, his foot caught in the hole, by which means he was thrown down and broke his thigh. It was proved the surgeon's bill amounted to 20 guineas. The jury found a verdict for the plaintiff with £40 damages.'

Robert Sibbald (1641–1722) was an illustrious gentleman, physician to Charles II, founder of Edinburgh's Royal College of Physicians and first Professor of Medicine at the University of Edinburgh. How great then his fall from dignity when his spurs locked together one day as he left his house and, tottering and slipping, he pitched headfirst onto the wet cobblestones.

A contemporary of Sibbald, equally illustrious, was Jean-Baptiste Lully, a musician in high favour with King Louis XIV. He used a long staff to beat the time on the floor and in 1687, while conducting a *Te Deum* to celebrate the royal recovery from a serious illness, he struck his foot with the staff instead of the floor. An abcess formed, gangrene set in and he died shortly afterwards. Perhaps, like Julie Adams at the start of this chapter, he had been wearing thin shoes.

As most of us have had drummed into our ears from an early age, fireworks can be very tricky things — hazardous to handlers and spectators alike, as a glance at some of the tombstone inscriptions will show. Experimenting with the mini-explosives can, of course, have equally painful consequences. The following report was published in London on 29th July 1795:

> 'Letters from Vienna state the unfortunate death of the Archduke Leopold, Palatine of Hungary, and brother to the Emperor, on Friday the 10th inst. On that day, the Emperor and his brother amused themselves at the Imperial Palace at Luxembourg, near Vienna, with preparing fireworks, assisted by a page and an Hungarian chasseur. The Emperor had been some time superintending this business, when finding the room warm, he walked out for the benefit of the air. The Archduke wished to try the effect of a rocket at one of the windows, but it rebounded back again, and set fire to the powder and other fireworks which were there. Everyone ran to lend all possible assistance as soon as the explosion was heard, but all efforts were in vain. The Archduke expired, after suffering fifteen hours of excruciating pain.'

One may well feel that the Archduke asked for it, but a victim of fireworks mentioned in Sir William Knighton's journal in 1831 was an innocent bystander:

> 'I may as well mention an anecdote of the landlady of the inn at Canterbury. During the late election, a rocket was thrown into the doorway, where she was standing with others. It glanced by the waiter's ear, and then rested on the eye of the unhappy young person, the landlady. The rocket immediately burst, and completely destroyed the eye. She underwent great torture, and in the hopes of being less disfigured, (for she was very handsome previously to the accident) she resolved to have a glass eye put in, which she now wears...'

THE FOLLY OF IGNORING WARNINGS given by those with superior knowledge is well illustrated by the following incident. In September 1886 it had been announced that there would be a great explosion on the following Saturday in the Crarae Quarries at Lochfyneside in Scotland, where a large quantity of rock was to be released prior to being dressed for paving slabs. Over 1,000 eager spectators travelled by the Clyde steamer *Lord of the Isles* to witness the operation. Fifty thousand tons of granite were loosened by the explosion of gunpowder. Visitors then clambered into the quarry, despite the efforts of a company employee to warn them of the sulphurous fumes which had not dispersed. Several of these intrepid spectators fell to the ground in convulsions, six died immediately, and one elderly gentleman was reported to have drawn his false teeth into his gullet and choked to death.

In 1836 the owner of a junk shop came by an old diving suit, complete with helmet. He was inquisitive as well as acquisitive, and his near-miss was reported at the time:-

> 'Mr. Caston determined to try it in the first instance on *terra firma* and for this purpose drew the helmet over his head, and then adjusted that part which fitted the lower extremities. He however omitted the most essential part of the apparatus, namely the valve that admitted the air into that portion which fitted over his head and face. This neglect nearly cost him his life; for when one of his servants entered the warehouse, Mr. Caston was discovered rolling on the floor, enveloped in the diving apparatus, apparently in great agony, and the servant saved his master's life by extricating him from the horrendous apparatus.'

A NEWSPAPER REPORT IN AUGUST 1777 caused distress in polite society:-

> 'The Right Hon. the Earl of Harcourt died, 16th Aug., 1777, at his seat at Newnham, in Oxfordshire. His Lordship had gone out to take his morning's walk in the park, and did not return at his usual hour, was found by his servants in a narrow well, nothing appearing above water but the feet and legs, occasioned (as it is imagined) by his over-reaching himself in order to save the life of his favourite dog, who was found in the well with him, standing on his master's feet.'

Accident records include many cases of people falling into excavated holes. *The Times* of lst March 1856 reported the case of Matthew Gladman of Lewes who met his death by falling into the soil of a water closet. It seems the covering boards of his outside lavatory had been removed preparatory to its being cleaned out the following day. The door was tied shut with a rope. Mr. Gladman, however, went out in the middle of the night, untied the rope and stepped right in.

Matthew Gladman would have had sympathy for a gentleman who died on 2nd January 1900, aged 65. The Birmingham Coroner held an inquest two days later, when it was adjudged that Mr. Thomas Hopkins died from 'chronic inflammation of the liver, and dropsy accelerated by wound in back from breaking of chamber utensil on which he sat.'

THE DIARIST JOHN EVELYN wrote at Venice in June 1645:-

'An honest merchant told me that one day walking in the plazza, he saw the fellow who kept the clock struck with the hammer so forcibly, as he was stooping his head near the bell to mend something amiss at the instant of striking, that, being stunned, he reeled over the battlements and broke his neck.'

This was at the clock tower in the Piazza of St. Mark. The famous bell tower of Venice, the Campanile, has a long history and took its present, final form in 1513. Nearly 100 metres high and fitted in 1793 with one of the earliest lightning conductors in Europe, it stood proud and seemingly impervious to all hazards. But on the 14th July 1902 it subsided into a gigantic heap of 14 million bricks. Age and some unwise internal alterations were blamed and evidence of instability had led the authorities to ban the playing of music in St. Mark's Square two weeks before the collapse. The only casualty was the caretaker's cat, but an eyewitness reported:-

'I will never forget that precise moment... the side facing the cathedral crumpled and split, and while the crowd let out an anguished cry and the dismal sound of crashing ruins settled over the square the enormous pinnacle of the belfry swung slowly two or three times from left to right ... and the colossus caved in on itself it buckled, buckled, buckled and collapsed. The earth trembled, a huge cloud of dust rose up and the golden angel plunged into it. The tragedy was finished.'

A less spectacular collapse is memorable only because the gravity of the law seems to have overcome the Law of Gravity. It happened at the Assizes held at East Grinstead in Sussex in March 1685, and again we have an eyewitness account:-

'... Upon a Tryal between my Lord Howard and another great person the Floor of the Nisi Prius Court fell down and with it all the Jury and Gentlemen Councele and Lawyers all into the Cellar ... By good Fortune or rather Gods Providence the Bench whereon the Judge sate fell not with it, but hung almost to a miracle.'

THE HAZARDS OF A RELIGIOUS LIFE are not easily identified. Nevertheless a task analysis of those spending much of their working day in and around churches must include, as we have just seen, the possibility of a collapsing bell tower. In this connection, an inherently safer system for church bells is demonstrated at the Church of Our Lady and St. Finnan, just outside Glenfinnan in Argyle. The church was built in 1870, evidently by a committee, for a bell was ordered although no bell tower had been incorporated in the design. The bell was therefore mounted just above ground level, where the faithful are invited to 'Ring and Pray'. However, visitors would be well advised to stand to one side when ringing and praying as the suspension looks none too sound. Alternatively, safety shoes are suggested in case the heavy bell falls the two or three inches onto your toes.

One place where a bell tower *was* provided for was at St. Mary's Church in East Bergholt, Suffolk. Building began in 1525 but was never completed. One reason given was the fall from power of Cardinal Wolsey; another view was that the merchant who had promised to fund the bell tower fell on hard times and defaulted. It would, however, be more satisfying to think that the church authorities appreciated the hazards of having some four tons of metal suspended over the heads of the bell ringers and put a stop to the building. At all events, a temporary ground level cage for the bells was erected in 1531, but was moved in the 17th Century because of noise pollution caused to a local landowner. The five bells are now rung by hand, the ringer, wearing ear defenders, standing alongside the bells and clearly safe from any failure of the bearings (renewed in 1972).

A wretched state of affairs at Old Shoreham in Sussex was revealed in 1675, when the Churchwardens plaintively reported to the Bishop of Chichester:-

> 'Church a great part of it is fallen downe and the rest in greate danger of falling, if some speedy corse be not taken to prevent it. The bells are all broken excepting one. The parsonage house hath bin for many yeares past totally ruined and fallen down.'

Another ecclesiastical danger is the possibility of being killed by a clock weight falling from the church clock tower. This happened, sadly, in 1756 to the Rev. Edward Moore, Curate of St. Chad's Church in Over, Cheshire, shortly after officiating at a wedding.

And not only falling clock weights. Matthew Paris, that 13th Century chronicler of monastic life, reported an unfortunate accident. He wrote in 1248 that 'a certain prior of the canons of a small church not far from the monastery of St. Albans was inspecting a pile of corn, commonly known as a rick, in order to estimate its value, when the rick, which was ill-constructed and leaning over, suddenly fell on him. Before the sheaves piled over him could be removed, this same prior, a plain fellow of little substance, died, miserably suffocated.'

A MEDIAEVAL FEMINIST must have been on an assertiveness training course and came by her just deserts, for a Coroner's Inquest reported:-

> 'On Wednesday before Ascension day 51 Henry III [25 May 1267] William de Stansgate came down a road called Burleyesdam with a cross-bow on his left shoulder and a poisoned arrow, and he met Desiderata, late the wife of Robert le Champeneys, who was his child's godmother, and a particular friend. And she asked him in jest whether he were one of the men who were going about the country with cross-bows, bows and other weapons, to apprehend robbers and evil-doers by the king's order; adding that she could overcome and take two or three like him. And putting out her arm she caught him by the neck and crooking her leg behind his, without his noticing it, she upset him and fell on him. And in falling she struck herself in the side with the arrow which he had under his belt, piercing to the heart, and died on the spot. Verdict. Death by misadventure.'

Another Coroner's Inquest, on the 29th January 1554, recorded an accident involving unguarded machinery:-

> 'At 2 o'clock, Amy Lewes, aged 13, servant of John Castilman of Chister, wandering idly in the city, came to a horsemill belonging to John Knott of Chister, then a miller, and went so carelessly within the sweep of the mill's arms, which were being turned by horses, that one of the arms struck her on the right side and killed her.'

BEFORE THE HAPPY ARRIVAL OF CENTRAL HEATING, pans of burning charcoal were frequently used to heat rooms that had no fireplaces. But it was a dangerous practice, and 'a very dreadfull instance of the deleterious effects of the fumes of lighted charcoal' was reported in March 1785. Mrs. Seymour was very ill, and her two sisters and a nurse sat with her. About midnight the temperature fell and some charcoal was lighted in a pan. The following morning Mr. Seymour entered the sickroom, 'when to his inexpressible horror, he found his wife, one of her sisters, and the nurse dead, and the other scarcely sensible. Medical assistance was instantly procured, but too late to recover any save the last person.'

They never learn. Here is a gruesome story of how, one hundred years later, they were still using charcoal to heat bedrooms and still dying as a result.

On 20th January 1881 a man, his wife and daughter went to bed at about 7 o'clock in a room containing a charcoal burning stove. At 5 o'clock the following morning the husband woke up and although in a dazed state was able to light a lamp. He must have been very dazed indeed because, although he found his daughter quite dead and his wife comatose, he lay down again beside his wife and fell asleep. He reawoke at 8 o'clock, belatedly realised his wife and daughter had been affected by charcoal fumes, and attempted to open the valve of the stove. He failed. He sat on the edge of the bed for a few hours, roused himself sufficiently to go out for a meal and a drink, and returned home without speaking to anyone.

That was on the 21st January. The following day he went out again and on returning home remained at the bedside until the morning of the 25th, by which time of course his wife was dead. He reported the two deaths and on being questioned said he thought his wife had died on the 24th. He was arrested and became so excited and troublesome that the police thought he was drunk, but after careful examination was found to be sober. It must be assumed that the charcoal fumes had affected whatever rational powers he had once possessed.

# CHAPTER 2

# AT WORK

WHEN CONSIDERING ACCIDENTS AT WORK, it is common to think in terms of factories or offices or building sites, but of course almost anywhere can be a place of work for somebody. The top of a ladder, for instance, is the workplace for a window cleaner and the hazards are obvious, while the stage at a rock concert is the workplace for a performer in danger of being electrocuted by his own guitar. Sometimes there is a curious kind of cross-fertilisation between worker and workplace. For five years, from 1905, Winchester Cathedral was the unlikely workplace for diver William Robert Walker as he battled, lead-weighted and brass-helmeted, to underpin the flooded and tottering foundations of the Cathedral. Nevertheless, most of the hazards and accidents noted in this chapter are located in traditional places of employment.

THE BUILDING SITE was probably the earliest formal workplace. The sketch on the following page, from the wall-painting in the tomb of the sculptor Ipy (1200 BC) at Deir el-Medina, Western Thebes, shows that even the ancient Egyptians thought accidents at work worth recording.

In the top right hand corner is an afflicted worker who looks as though he is having a dislocated shoulder re-set by a first-aider, while the hairless little fellow on the lefthand scaffolding has clearly dropped a mallet on his big toe. Below him another first-aider is removing something from a worker's eye. The lad in the bottom righthand corner is on a tea break.

The construction industry has had its quota of hazards and the changeover from Romanesque to Gothic architecture in the Middle Ages brought special problems. The roof of Beauvais Cathedral fell in twice in the 13th Century, while the chapter house of the great Batalha monastery in Portugal also collapsed twice during construction and was finally built by prisoners already condemned to death.

THERE SEEMS ALWAYS TO HAVE BEEN an awareness of the dangers workers were exposed to, although until comparatively recently little

was done to protect them†. In the Roman Empire, for example, the life of a slave working in the lead industry was reckoned in months rather than years, so great was the likelihood of lead poisoning in the cupellation of lead ores for the recovery of silver.

AT ONE TIME A PARTICULARLY DISTRESSING fate† awaited the inattentive merchant seaman, and among the regulations issued in 1553 by Sebastian Cabot, Governor of the Company of Merchant Adventurers, we read the following:-

> 'Item, there are people that can swimme in the sea, havens & rivers, naked, having bowes and shafts, coveting to draw nigh your ships, which if they shal finde not wel watched, or warded, they will assault, desirous of the bodies of men, which they covet for meate; if you resist them, they dive, and so will flee, and therefore diligent watch is to be kept both day & night, in some Islands.'

†The UK can pat itself on the back for having introduced the world's first factory legislation. This was the Health and Morals of Apprentices Act. It was passed in 1802 and made certain provisions for cleanliness and ventilation.

DIARIST JOHN EVELYN came across an environmental hazard in Rome. Writing in May 1645 about a visit to an asylum, he noted:-

'The garden of simples is well furnished, and has in it the deadly yew or *taxus*, of the ancients; which Dr. Belluccio, the superintendent, affirms that his workmen cannot endure to clip for above the space of half an hour at a time, from the pain of the head which surprises them.'

ON ONE OF HER FAMOUS JOURNEYS Celia Fiennes went to Derbyshire in 1697. She reported on the dangerous conditions the miners worked under:-

'Thence to Buxton, over those craggy hills whose bowells are full of Mines of all kinds of black and white lead and veined marbles, and some have mines of Copper, others Tinn and Leaden mines in which is a great deal of Silver; ... they digg down their mines like a well, for one man to be let down with a rope and pulley and so when they find oar they keep digging underground to follow the oar, which lies amongst the stone that lookes like our fire stones; in that mine I saw there was 3 or 4 at work and all let down thro' the well, they digg sometimes a great way before they come to oar; ... they wall round the wells to the mines to secure their mold'ring in upon them; they generally look very pale and yellow that work Underground, they are forc'd to keep lights with them and some tymes are forced to use Gunpowder to break the stones, and that is sometymes hazardous to the people and destroys them at the work.'

Another traveller, John Woolman, Quaker and reformer, was struck by the hardships endured by post boys employed on stage coaches. In 1772 he lamented in his diary:-

'Post boys pursue their business, each one to his Stage, all night through the winter. Some boys who ride long Stages suffer greatly in winter nights and at several places I have heard of their being froze to death.'

and he added a comment we hear today, although not usually expressed in such florid terms: 'So great is the hurry in the Spirit of this world, that in aiming to do business quick, and to gain wealth, the Creation at this day doth loudly groan!'

ALL IN ALL, THE WORKERS HAD A PRETTY ROTTEN TIME OF IT in the 18th and earlier centuries. *The Gentleman's Magazine* summed it up in 1782:-

> 'The collier, the clothier, the painter, the gilder, the miner, the makers of glass, the workers in iron, tin, lead, copper, while they minister to our necessities or please our tastes and fancies, are impairing their health and shortening their days... suffocated in mines and pits, or gradually poisoned by the noxious effluvia of metals, oils, powders, spirits, etc., used in their work, and can exhibit as mournful a scene of blinded and lame, of enfeebled, decrepit, asthmatic, consumptive wretches, panting for breath and crawling half alive upon the surface of the earth.'

IT IS HARD TO ASSOCIATE the working conditions described above with the sedate peace and security of a classroom. Certainly in the good old days of school discipline (as we like to think) demure children respectfully absorbing the wisdom of their elders hardly represented a comparable hazard. But is this so? In 1914 a schoolteacher was assaulted by some of his pupils in a preconceived plan and was so severely injured that he died. The courts had to decide whether he had met his death by accident in pursuance of the duties of his employment. The judge ruled that he had and awarded in favour of his mother.

MORE RECENTLY A KITCHEN HAND at a restaurant had a distressing accident. It was a hot summer day and, thinking to get some fresh air, he stood on a stool to open the window. He reached forward and to balance himself stretched one leg out behind him. Unfortunately, though, he was wearing only sandals on his feet and the bacon slicer was still in operation. In stretching too far his big toe was sliced off.

There was a more celebrated loss of limb in 1737. Samuel Wood was twenty-six years old at the time and was employed at a mill near the Isle of Dogs. He became entangled in the cogs of one of the large wheels and his arm and shoulder blade were torn from his body. A contemporary account reports that 'At the Time the Accident happen'd, he says he was not sensible of any Pain, but only a tingling about the Wound, and being a good deal surpriz'd, did not know that his arm was torn off, till he saw it in the wheel.' First aid was to hand, and somebody '... strew'd a large quantity of Loaf-Sugar powder'd into the Wound, in order to choak the Blood.'

Samuel Wood (the miller).

Medical assistance was sought and so incoherent a message did the doctor receive that he arrived on the scene with an apparatus for setting a broken arm but with no dressings (a splendid example of the hazards of giving imprecise information when calling in the emergency services). The doctor sent for suitable equipment, stuffed the exposed blood vessels back into the wound, stitched it up '... by means of a Needle and Ligature' and applied a dressing. Samuel was taken the next day to St. Thomas's Hospital, where the dressing remained undisturbed for four days, as a result of which benign neglect the wound healed itself.

He survived for more than twenty-five years as a Customs official and publican with no apparent post-traumatic stress syndrome. In the absence of a disability allowance he resorted to the eighteenth century equivalent of selling his story to the press and augmented his pay by the sales of a print depicting the accident in grisly detail.

So remarkable a recovery was Samuel Wood's case thought to be that an interested surgeon gave an account of it to the Royal Society and concluded his address: '... and for the farther satisfaction of the Society I have brought the Man himself, and likewise the Arm, just as it was torn from his Body, which has been kept in Spirits ever since the Accident happen'd.'

The publicity surrounding this accident seems to have been concentrated solely on the treatment of Wood's injuries, with no attention being paid to their cause. Hence it should not surprise us when, fifty years later in Liverpool, history repeated itself. As reported with unseemly gusto by the *New Annual Register* on 6th September 1790:-

> 'On Friday morning, about five o'clock, George Drover, who takes care of a mill, near Limekiln-lane, belonging to Messrs. Pennington and Part, went into the lowest apartment, in order to grease the wheels, without having the precaution to stop the mill; in doing which the cogs unfortunately caught the fingers of his right hand, drew in his arm, which it broke all the way up, when his chest, being by that means forced against a frame, through which the wheel in its operation passed, and which being too small to admit his body, the arm was torn from it about four inches below the shoulder. In this dreadful situation, which was accompanied at first by a great effusion of blood, he went up six or seven steps which led to the bank on the outside, to stop the mill, and sat some time before he called to some ropemakers in the neighbourhood, who came to his assistance; when having procured a chair, they carried him to our Infirmary, where his arm was amputated, and the remaining stump taken away from the shoulder-joint by Mr. Park, and yesterday evening had every appearance of a cure.'

THE INDUSTRIAL REVOLUTION was gathering momentum and with it the frequency of such accidents. Also growing was public anxiety, as the coroner's verdict on the following death shows. A press report from Sheffield, dated 2nd August 1792, told the story.

> 'A dreadful accident happened, on Friday, at the steam-wheel in Green-lane; John Smith, an iron-founder, imprudently ventured too near the interior works to examine them, when the cogs of one of the wheels caught hold of him, and a shocking, though happily for the sufferer, an instantaneous death was the consequence. A

coroner's inquest was taken, and the jury brought in their verdict accidental death; they also levied a fine of £5 upon the proprietors, intended to operate as a caution to owners of such works how they admit persons to inspect them without a guide.'

The Association for the Mangling of Operatives, incidentally, was the label Charles Dickens applied to the Factory Law Amendment Association, a group set up by mill owners to lobby for the repeal of those provisions of the 1844 Factories Act that dealt with the fencing in of dangerous machinery.

But such accidents continued. Allenheads in the north east of England was once the largest silver-lead mine in the world and had installed a water-wheel inside the mine to improve ventilation. In 1879 Thomas Heslop got entangled in the wheel. His body was literally torn to shreds which, to the horror of bystanders, were washed out through the entrance to the mine, piece by bloody piece.

MINING HAS ALWAYS BEEN ONE of the more dangerous occupations, and the great disasters are well known. But lesser disasters in mines and other areas of activity could devastate a town or village. Such a tragedy is recorded in the village of Little Dean in the Forest of Dean, Gloucestershire:-

'These four youths were suddenly called into eternity on Tuesday the 6th day of April 1819 by an awful dispensation of the Almighty. The link of a chain employed to lower them into Bilston Pit breaking they were precipitated to the bottom of the Pit. Their bones literally dashed to pieces their bodies thus presenting a frightful and appalling spectacle to all who beheld them. They were interred in one grave on the Friday following being Good Friday April 9 1819. A Funeral Sermon was preached on the mournful occasion on Sunday April 25 1819 in the Church of Little Deane before a congregation of 2500 people on the following text which it was judged it was advisable to record on their Tomb stone as a suitable admonition for the benefit of all survivors.

Luke XIII vs 1,2,& 3.2

'Swift flew the appointed messenger of death and in a moment stopt their mortal breath. Art THOU prepared as suddenly to die?

Tis mercy's call O list unto the cry.
Thomas Morgan Aged 26 William Tingle Aged 19
Robert Tingle Aged 16 James Meredith Aged 12'

There have also been countless accidents to individual miners, and the Rev. John Skinner, Rector of Camerton in Somerset in the last century, had ample scope for exercising pastoral care of those who worked in the coal pit in his parish. His diary is replete with sad little stories, to which he could not restrain himself from adding a censorious gloss. On 9th January 1806 he wrote:-

'William Brittain, another collier, was killed in the coal pits by a shocking accident. He was riding in a small coal cart underground drawn by an ass... As the ass was going along at a brisk pace he did not observe he was come to a spot where the roof of the passage was much lower than it was before, and, neglecting to stoop his head, his back was bent double by the sudden violence of the shock and his spine snapped. The poor fellow was drawn up and lived some hours, but his extremities were quite paralysed. He was son to poor old Britain who occasionally works for me. The brother who is left will be of little comfort to him, as he is a sad fellow, indeed quite a savage.'

and on 22nd March of that year another parishioner was killed:-

'James Edwards, whose business it was to see the coal brought to land at The Old Pit, in reaching over too far in order to stay the basket which was coming up, fell to the bottom and was dashed to pieces. Horrid to say, his last word was an oath when he found himself going.'

On 20th August 1824 he was disappointed at not finding a grieving widow:-

'Having heard that young Tyler was killed by a stone falling on him in the Coal Pit last night, I went to the house and found his body had been brought home and was laid in the cottage, but was very properly covered up, as it was much disfigured. His wife, who is far advanced in pregnancy, did not seem much afflicted.'

A BIZARRE ACCIDENT HAPPENED in the 1940s at an acid concentration plant. Fortunately it had a happy ending. Because of acid leaks at such plants, it was customary for employees to wear old raincoats and hats, but invariably the outer waterproofing disintegrated before the woollen lining, and if acid did splash on to the skin it could be quickly washed off at the nearest water tap. However, before emergency showers were available, more seriously acid-splashed workers had recourse to one of the enamel baths of water dotted around the plant. On one occasion there was a serious leak at the plant and an operator was badly covered in acid. He ran to the nearest bath and jumped in. But it was mid-winter and the bath was solid with ice. He skidded across it, shot off the other end and broke his leg. Fellow workers managed to wash him down with water but he was seriously burned.

He recovered from the burns and the broken leg, and married the nurse who had treated him at the plant's medical centre.

The factory building itself precipitated a sad calamity in 1792. As reported from Macclesfield on 23rd November:-

> 'This day, about five minutes before twelve o'clock, a dreadful accident happened at a cotton manufactory belonging to Messrs. Clayton and Gaskill, in this town. A great part of the roof of that extensive building fell in, while all the hands were at work, owing to the timber's drawing from the walls. A great number of persons are buried in the ruins. Several persons have been taken out dead, and many alive but greatly bruised. It is supposed that about 50 or 60 men, women and children were under the roof at the instant when it fell in. Only 16 are found. The cries of those buried are exceedingly distressing. All possible means are using to liberate the living, and to dig out the dead. One part of the front wall was apparently falling every moment; which prevented the populace from giving any assistance for upward of two hours. The wind being extremely high, threatened every moment to blow down an adjacent part of the building. Several thousand of people were assembled on the alarm. A brave Irishman, at the hazard of his life, was determined to liberate two men whom he perceived crying out for help, which was soon afforded them. The populace, animated by his example, lent their assistance. Every surgeon in town cheerfully rendered his best services. One man and one child were found with their heads severed from their bodies, and bruised in a shocking manner.'

The spectacle of doctors working cheerfully on such an occasion must have been a touch macabre.

GRACE DARLING'S STORY is well known and is a reminder of the dangerous nature of work on a lighthouse. The saga of Eddystone is worth repeating.

The Eddystone rocks off Plymouth were always a hazard to shipping and in 1696 Henry Winstanley, nephew of Charles II's clerk of works, undertook to build a lighthouse on them. The first light shone forth on 14th November 1698 and in the following Spring Winstanley, after surveying winter damage, added some strengthening building works. They were evidently not strong enough, for in November 1703 there was an exceptionally violent gale and the lighthouse was swept away, together with Winstanley himself.

In 1706 another attempt to build on Eddystone rocks was made by John Rudyerd, and his lighthouse lasted until 2nd December 1755 when it burned down. The fire, not surprisingly, started in the lantern and seems to have been caused by carbon and grease smouldering from a crack in the chimney of the kitchen stove, which ignited when the 94 year old keeper went to check on the light. He was unable to put out the fire and his two 'assistants' did nothing to aid him, despite his calls for help. The Superintendent of Plymouth Dockyard wrote an account of the fire on 12th December, and speculated on its causes:-

> '... they remained on the Rocks with the Melted Lead and pieces of Burning Timber frequently falling upon them, till about three of the next afternoon, when some men in a Cawsand Boat got them off.
>
> 'Whether this Accident proceeded from their Drinking too much of three Pints of Gin received with the Provisions and Candles that Evening, or having too great a fire in the Chimney, I cannot discover, but must observe at the time the Fire was discovered there was but one Buckett in the House and what I think very improper they have never been allowed small Candles for the necessary uses about the House, but have always burnt the Ends of Large ones that were taken from the Lantern and stuck on a piece of Wood between three Nails, for a Candlestick which is certainly attended with great Danger ...'

The geriatric keeper came to a harrowing end. Once on shore he managed to claim that he had swallowed molten lead. He died twelve days later and an autopsy indeed revealed a 7 oz lump of lead in his stomach. The keeper lives on in history, however; the lead ingot is now in the National Museum of Scotland in Edinburgh.

A WELDER SUFFERED an equally horrific death more recently. He had great difficulty in lighting the burner and smelt the tip to see whether acetylene was coming through. When the gas mixture was eventually lit, the acetylene–oxygen mixture that had already escaped exploded, and the explosion extended to the gas that had entered the welder's mouth and lungs when he sniffed the tip. He died as a result of his injuries. They included burns to the mouth, throat, internal walls of the wind pipe and bronchial tubes; in addition the pleura had been torn by the force of the explosion.

In 1866 John Ruskin published *The Crown of Wild Olive.* In a footnote to his Preface he tells his readers:-

> 'A fearful occurrence took place a few days since, near Wolverhampton. Thomas Snape, aged nineteen, was on duty as the 'keeper' of a blast furnace at Deepfield, assisted by John Gardner, aged eighteen, and Joseph Swift, aged, thirty-seven. The furnace contained four tons of molten iron, and an equal amount of cinders, and ought to have been run out at 7.30 pm. But Snape and his mates, engaged in talking and drinking, neglected their duty, and, in the meantime the iron rose in the furnace until it reached a pipe wherein water was contained. Just as the men had stripped, and were proceeding to tap the furnace, the water in the pipe, converted into steam, burst down its front and let loose on them the molten metal, which instantaneously consumed Gardner; Snape, terribly burnt, and mad with pain, leaped into the canal and then ran home and fell dead on the threshold; Swift survived to reach the hospital, where he died too.'

RAILWAY ACCIDENTS involving passengers are dealt with in Chapter 4, but what about the navvies who built them? Accidents there were aplenty before a single passenger bought a ticket, for the workmen laboured not only at ground level but also in dangerous tunnelling operations. Sometimes negligence above caused death below. It is reported that in 1839 a worker was returning a waggon to the shafthead of the Bishopston Tunnel on the Glasgow, Paisley and Greenock line and, not looking where he was going, pushed the waggon down the shaft where it crushed the head of a man at the bottom. The worker, apparently in tears, made a quick get-away from the scene but was later arrested in Glasgow.

Giving evidence to a Select Committee on Railway Labourers in 1846, the former secretary of the South Western Railway referred to a tragic but pointless mishap on the Paris to Rouen line in France:-

> '... in a tunnel they were using powder to blast a rock, and they made a hole, as is the practice; they primed it, inserted the powder, and expected it to go off; it did not go off, and a foolish Irishman was silly enough positively to go and blow. Two men were very severely hurt; the man who blew the powder had both his eyes blown out, and both his arms blown off.'

Many employees will recognise the sentiments expressed in William Louis Page's complaint a few weeks before he died in what was headlined as a 'Sad Accident at the Drainage Works'. As reported in the *Eastbourne Chronicle* on 1st August 1896:-

> 'The Coroner sat with a jury at the Rising Sun Coffee Tavern, Seaside, on Wednesday, to inquire into the circumstances attending the death of William Louis Page, aged 41, who was killed whilst employed at the new sewer outfall at Langney Point early on Tuesday morning. The deceased was engaged in the trench where the outfall pipes were being laid, when a timber brace, being accidentally dislodged from the gantry used in the process of lowering the pipes, fell upon the unfortunate man's head, causing injuries from which he died shortly afterwards.'

Mrs. Page, the victim's widow, said in her evidence that 'her husband complained to her about a fortnight ago that the man who had the management of the works did not know so much about it as he did himself'.

Canny workers in 1812 also seem to have known a thing or two about the project they were engaged upon. Parliament had authorised the driving of a tunnel under Highgate Hill to obviate the 1 in 10 gradient on this stretch of the main road from London to the North. (It was at the foot of this hill, incidently, that Dick Whittington heard the call to return to London and become Lord Mayor.)

The tunnel was to be 211 yards long and from the start doubts were being expressed about the soundness of the brickwork lining. About 130 yards had been constructed when, on the 13th April 1812, the tunnel collapsed. Happily, nobody was injured and, as one newspaper report stated, 'How providential that the fall was reserved for a moment when no person was on the spot, to suffer by an accident which has reduced this Herculean task to a heap of ruins.' But another journal, *The Gentleman's*

*Magazine*, had evidently interviewed some of the construction workers and commented:-

> 'The falling-in of the archway had been anticipated by the workmen for nearly a fortnight previous to the catastrophe, and is considered to have originated in too economical a regard to the quantity of bricks used in the arch, and the quality of the cement for uniting them having been deteriorated by too great a proportion of sand and lime...'

The tunnel project was abandoned and an open cutting adopted with an arch spanning it. This road is now called the Archway.

MEDIA SENSATION MONGERS were hard at it in the 18th Century, and in reporting in the national press a tragedy that took place on 23rd October 1799 the finger of blame was pointed at irresponsible (and macho) workmen:-

> 'A dreadful accident happened at Coalport, near the iron bridge, Shropshire, this evening. As a considerable number of the work-people, men and women, belonging to Messrs. Rose and Co.'s china-works at that place were returning from the manufactory to their homes on the other side of the Severn, in a large boat kept for that purpose, some of the party very imprudently rocking the boat in a very violent manner, in order to intimidate the women, the melancholy consequence was that too great a number crowded towards the head of the boat, which took the water, and the greater part of the persons on board were precipitated into the stream; when twenty men and eight women lost their lives.'

But the owner of the pottery, Mr. Rose, wrote to the *Salopian Journal* a few days after the accident:-

> 'As there will be many erroneous accounts circulated respecting the unfortunate and shocking accident that happened here on Wednesday last evening at 9 o'clock at night, I beg leave to state as near the particulars as yet I have been able to learn.
>
> 'As the people from the Coalport Manufactory, to the number of 43, were leaving their work at 9 o'clock at night, to go home over the usual passage boat, owing to the inattention of the man whom the boatman had entrusted to steer over, the boat unfortunately went down with all on board, when only 15 out of the whole

could save themselves; the remaining 28 were unfortunately lost. In consequence of the great fog, and darkness of the night, no-one was able to give the least assistance. Ten have since been taken out of the water, and the Coroners inquest is to be held over them this day.'

**CARRYING BUCKETS, BOTTLES AND SIMILAR CONTAINERS** is only hazardous if they contain dangerous substances (we can ignore Jack and Jill for the moment). The causes ascribed to the following accidents could not, of course, be verified by the person carrying the container but the precarious nature of this type of equipment should be noted by reliability engineers.

- In March 1870 at Paterson, New Jersey in the USA, a man dropped a can of nitroglycerine he was carrying. The explosion killed him and caused another explosion in the store house in which several other people were killed.
- In November 1950 at Maribyrnong, in Australia a man was killed in an explosion caused, it is supposed, when he dropped a rubber bucket containing 30 lb of nitroglycerine.
- In January 1954 at Schlebusch, Germany, three men were killed when, it is believed, one of them stumbled while carrying a container of nitroglycerine.
- In April 1961, again at Schlebusch, it is thought that an aluminium bucket containing an explosive was dropped, killing two men.
- In February 1962 at Penrhyndendraeth, Wales, a hard rubber bottle of nitroglycerine was dropped on the floor and killed the man carrying it.
- In March 1963 at Modderfortein, South Africa, a bucket of nitroglycerine was the probable cause of an explosion which detonated some 14 tons of the explosive.

The list could go on and on. Plainly this is no job for the butter-fingered employee.

**PROTECTIVE CLOTHING FOR THE WORKFORCE** has come a long way since 18th century firemen were kitted out in the livery of the insurance companies who employed them. And very smart they were too, with knee breeches, knitted stockings and buckled shoes. An elegant outfit, however, is not the prime concern of a body of men engaged in dangerous activity and in 1763 a petition was addressed to the London Assurance pointing out that:-

> '... amongst all the Dangers and hardships to which Your poor Petitioners are exposed in the discharge of the Duties of their Office, there is none more sensibly felt than those which they experience from the want of Boots, as their Leggs are frequently torn with Nails, Barrs of Iron, and such kind of Rubbish as Fires occasion.
>
> 'That the late dreadful Fire at Shadwell particularly Evinced the great Necessity of Boots, as several of Your poor Petitioners were up to their Knees in Water, hot from the Water-Works, and Instantly after plunged in Cold Water, by which Deplorable case great Numbers of Your Petitioners Lives were endanger'd, by the Coughs and Colds which they caught, which Calamity, Your Petitioners presume to suppose, might be in future happily prevented were their Leggs defended by Boots...'

A fireman's lot is generally not a happy one and in the last century industrialisation brought new risks for this honourable profession to face. In 1890 two firemen, called to a fire in Newcastle, were killed by breathing nitric acid fumes. The rest of the crew returned to their station after the fire was out and tried to clear their lungs by walking about in the fresh air of the station yard. Their common sense was frustrated when a doctor was called and made them inhale ammonia as an antidote, with the unfortunate consequence that they all collapsed. One of them died and their Superintendent was invalided out of the service the following year.

When they weren't being suffocated they were being electrocuted. Since 1866 British firemen had worn brass helmets, copied from the French *Sapeurs-Pompiers* who had copied them from the military. They were discarded in 1937 after too many men had been killed on contact with bared electrical wiring, and were replaced by compressed cork helmets safe to 15,000 volts.

VATS AND SUCH seem to be dangerous pieces of equipment to have around and the risks of falling into them need no elaboration. A distressing accident of this kind happened in a place called Boone's Lick, Missouri, where this notice appeared:-

> 'Tragedy.
>
> A great tragedy occurred on August 10, 1833, when James Morrison's 16 year old son, Joseph, fell into one of the boiling

kettles and was horribly scalded. After much suffering, he died and was buried nearby. James Morrison never returned to Boone's Lick; he sold it the following year.'

Here are some more accidents from the other side of the Atlantic. Considering the contents of the vats, we cannot help wondering whether the victims were hungry as well as foolhardy:-

- A worker in Downington, Pennsylvania, removed the lid of a vat of chocolate. He fell in and was killed. The rescuers had to cut into the sides of the steel vat to remove his body.
- In Warren, Michigan, an employee died when he fell into a vat of gravy being prepared for a chain of restaurants.
- In El Dorado, Arkansas, a plant worker entered a vat containing chopped chicken to free a blocked valve. He was wearing a gas mask but detected the smell of ammonia and started to climb out. He fell back into the vat. Another worker went in to help, probably without a gas mask, and was also overcome by the gases. Both workers died.c
- And in Toronto a worker fell into a vat of pizza dough and was asphyxiated.

In Deer Park, New York, it was the vat that fell over the worker and killed him. It contained molten lipstick.

But the incident to cap all these vat-related misadventures took place in France where, on 22nd January 1837, it was related that:-

'A man belonging to the beet-root sugar manufactory of St. Saulon, near Valenciennes, a few days ago, in a moment of irritation arising out of a dispute, threw one of his companions into a vat full of syrup. The victim scrambled out, and hastened to the town to lay a complaint before a magistrate. The frost was very severe; and before he reached his destination the syrup became so completely candied that he bore all the appearance of an enormous stick of barley-sugar, insomuch that when he came to the magistrate's door, his arms were glued to his sides, and he was obliged to entreat a person passing by to pull the bell for him. The plaintiff received 15fr. damages for the assault, and paid 20fr. to the proprietors for loss on the syrup.'

# CHAPTER 3

# AND AT PLAY

SOCIOLOGISTS HAVE BEEN WARNING US that with robots and computers replacing the human workforce we will all have a problem of how to fill up our increased leisure time. This, of course, is a pseudo-proposition as the problem for most of us is how to get enough leisure to devote to the enormous array of activities now on offer. For leisure is indeed the new growth industry. Many of its products — scuba diving, hang gliding, pot holing, to name a few — are of their nature hazardous pastimes and those who enjoy living dangerously must take their chances. However, the less obviously death-defying ways of relaxing outside factory and office are also beset with surprising risks.

**THEATRE-GOING,** for instance, has had more than its fair share of calamities over the years. Let us contemplate Act V, Scene 1, of Christopher Marlowe's *Tamburlaine the Great*. The Governor of Babylon appears hanging in chains on the walls and Tamburlaine is speaking:-

*No, though Asphaltis' lake were liquid gold,*
*And offer'd me as ransom for thy life,*
*Yet shouldst thou die. — Shoot at him all at once.*
*[They shoot]*
*So, now he hangs like Bagdet's governor,*
*Having as many bullets in his flesh*
*As there be breaches in her batter'd wall.*

Stirring drama, no doubt, but probably not appreciated by all members of the audience at a performance in 1587. Giving a new depth to the meaning of 'live theatre', one of the armaments was loaded with real shot. As described by one Philip Gawdy in a letter to his family on 15th November of that year, an actor 'missed the fellowe he aymed at and killed a chyld and a women great with chyld fortwith and hurt an other man in the head very sore.'

The Elizabethans obviously regarded places of public entertainment as dens of all manner of iniquities — yea, even to the mayhemings of the queen's subjects — hazardous to body and soul. The preamble to an Act of the Common Council of London in 1574 notes that:-

> Sundry great disorders and inconveniences have been found to ensue to this city by the inordinate haunting of great multitudes of people, especially youth, to plays, interludes, and shows — namely, occasions of frays and quarrels; evil practices of incontinency in great inns having chambers and secret places to their open stages and galleries; inveigling and alluring of maids, specially orphans and good citizens' children under age, to privy and unmeet contracts; the publishing of unchaste, uncomely and unshamefast speeches and doings; withdrawing of the queen's majesty's subjects from divine service on Sundays and holidays, at which times such plays were chiefly used; unthrifty waste of the money of the poor and fond persons; sundry robberies by picking and cutting of purses; uttering of popular, busy and seditious matters; and many other corruptions of youth and other enormities — beside that also sundry slaughters and mayhemings of the queen's subjects have happened by ruins of scaffolds, frames, and stages, and by engines, weapons, and powder used in plays.

A letter from Sir Henry Wotton to his nephew tells us that at the burning down of the Globe Theatre in 1613 the only casualty beyond 'a few foresaken cloaks' was a man who had his breeches set on fire, but 'by the benefit of a provident wit' he put out the fire with a bottle of ale.

Another theatrical disaster occurred on 8th September 1788 at St. Omer near Calais in northern France where there was a camp for 40,000 men. The play-house, built on the heath for the purpose of the camp, fell down during the presentation of *Richard Coeur de Lion*. The failure of the building caused the death of thirty people and seriously injured one hundred and fifty people. Among the injured was the principal actress who had her back broken in two places. The builder of the theatre was instantly seized and conveyed to prison (no fiddling around in those days with public enquiries or protracted litigation to establish culpability).

**IN JUNE 1785** it was a performer and not the spectators who fared badly. As reported, the busking acrobat's death was attributed to his greed.

> 'At Axminister revel, in Devonshire, held on Monday the 6th, a man (who had formerly acted in the capacity of a Merry Andrew) had the presumption to stand twice on his head on one of the battlements of the Tower, to the astonishment of a number of spectators; and being liberally rewarded for so doing, was

> induced to make a third attempt, in the performance of which he fell down, and was killed on the spot.'

A star performer of greater renown who died while entertaining her public was Madeleine-Sophie Blanchard, wife of the Channel-crossing pioneer. On the evening of 6th May 1819 she ascended in a balloon from the Tivoli Gardens in Paris to give a firework display. The band played and the enthusiastic crowds watched with delight as blazes of Bengal fire and golden rain showered down. Mme Blanchard then entranced her audience by sending down a bomb of silver rain swinging from a parachute, and when a burst of flame appeared a moment later the spectators thought it all part of the fun. But it was the balloon itself that was on fire, the hydrogen with which it was filled having been ignited by the portfire used to light the fireworks.

The balloon descended rapidly, with the intrepid Mme Blanchard hurling out ballast as it swept low over buildings. To no avail, for the car of the balloon struck a roof and pitched Mme Blanchard into the street where she broke her neck and died.

PLAYING IN THE STREETS has always presented dangers for children. This report from the Calendar of the Coroners Rolls for 1301 records:-

> 'On Tuesday the feasts of Sts. Philip and James a certain Hugh Picard was riding a white horse after the hours of vespers, when Petronilla, daughter of William de Wyntonia, aged three years, was playing in the street; and the horse, being strong, quickly carried Hugh against his will over Petronella so that it struck her on her right side with its right forefoot. Petronella lingered untill the next day, when she died, at the hours of vespers, from the blow. Being asked who were present, the jurors know only of those mentioned. The corpse viewed, the right side of which appeared blue and badly bruised, and no other hurt. The horse, valued at a mark, for which Richard de Caumpes, the sheriff, will answer. Hugh fled and has no chattels: he afterwards surrendered to John de Boreford, sheriff.'

But there is another activity that children would be well advised to avoid, and that is the family holiday, with the car breaking down, luggage being stolen, and the kids themselves falling ill. Back in 1794, Sir William Pepys (who shared an ancestor with the famous Samuel) knew all about it. He mentioned a few of the problems in a letter to his eldest son:-

'You will be glad to hear that we all arriv'd safe in Town, especially when I tell you that two of our Post Horses fell down at once in coming down a Hill on the Downs, and that, had it not been for the vigilance of little Sophy, whom I had plac'd to watch the Trunk behind, it would have been stolen, as five men beset it in Cockspur Street, One of Whom had got so far as to cut the Cord, but my Stentorian Voice soon put them to flight. We are therefore all (thank God) once more arriv'd safe within 20 miles of You. When I take out with Me a Wife and six Children into the Country (where no Assistance can be had in case of Illness) I cannot help feeling very thankful when I see them all return, not only well, but much improved in Health...'

EATING AND DRINKING are among the more pleasurable ways of passing the time. Thomas May, the 17th century poet and historian, apparently did both to excess and died in a fitting manner. As noted in Aubrey's *Brief Lives*, he:-

'Came to his death after drinking with his chin tyed with his cap (being fatt); suffocated.'

However enjoyable a convivial drink with friends may be, it is undeniable that alcohol is the root cause of too many accidents. Sometimes, though, the alcohol itself is dangerous. In 1930 some 15,000 people were affected by drinking an alcoholic beverage called Jamaica Ginger or jake. Ten of them died. The drink had been adulterated with about 2% of tritolyl phosphate, a highly toxic substance. This sort of thing is nothing new, for in 1810 it was reported that a solution of lead in acid gave a sweet taste to the liquors with which it was mixed. Hence lead acetate, also known as sugar of lead, was employed by ignorant or unprincipled dealers to correct the acidity of their wines and ciders.

This dire additive caused the deformity recorded by Thomas Raikes, himself somewhat of a *bon viveur*, in his diary for 1st October 1837:-

'General Edmund Phipps is dead at Venice, aged seventy-seven. He was a worthy good-natured man, and he had for many years lost the use of his right arm from paralysis, caused, as it is said, by drinking bad sherry when in regimental quarters, in which white lead was infused. This arm hung down like the fin of a turtle, which gave him the sobriquet of Governer of Finland.'

As for dining out, invitations should be carefully sifted to weed out (pun intended) the home grown vegetable enthusiast. The Rev. William Cole, Rector of Blecheley, jotted down in his diary on 1st March 1766:-

> 'Fine Day. Message from Mr. Holt of Loughton not to dine with him to Day, as his Tenent's Wife, Mrs. Yorke, died by eating Hemlock Roots instead of Parsnips & was to be buried this Evening from the House. Six others in the Family were poisoned also, & very ill, but got over it.'

Nor were Medieval banquets all they were cracked up to be if we can believe Peter of Blois, who wrote in 1160:-

> 'I often wonder how anyone... can endure the annoyances of a court life... The servants care nothing whatever whether the unlucky guests become ill or die... Indeed the tables are sometimes filled with putrid food, and were it not for the fact that those that eat it indulge in powerfull exercises, many more deaths would result from it. But if the courtiers cannot have exercise (which is the case of the court for a time in town) some are always left behind at death's door...'

THE POWERFUL EXERCISES advocated by Peter of Blois probably did not include cricket, although the courtly knights in their armour would hardly have felt out of place opposing a side of today's players, with their helmets and padding. Possibly because of this protective gear, even first class cricket is not considered a dangerous game, but here is a curiosity to add to the lore of the game.

The investigation into the death of a groundsman at a cricket ground revealed a novel use of electricity. To eradicate worms from the wicket area, this gentleman devised an apparatus consisting of a metal-tined garden fork and a length of single insulated flexible cable terminating in a crocodile clip at one end. The other end of the cable was connected to the live terminal of a 13 amp plug. The method of use was to insert the fork into the wicket area, attach the crocodile clip to the metal tines, plug in at a 13 amp socket outlet and energise the system. The full supply of 240 volts caused the worms in the region of the fork to rise to the surface where they would be collected. The groundsman died as a result of touching the crocodile clip while it was still energised.

We do not know what happened to the worms. Perhaps we can say, as a switch on Oliver Goldsmith's mad dog:-

*The worms recover'd of their fright,*
*The man it was who died.*

Football is a somewhat more forceful game than cricket, and the hooligans were evidently around in 1583 when Philip Stubbes wrote *The Anatomie of Abuses*:-

> 'For as concerning football playing, I protest unto you it may rather be called... a bloody and murdering practice, than a fellowly sport or pastime... sometimes their necks are broken, sometimes their backs, sometimes their legs... sometimes their noses gush out with blood, sometimes their eyes start out... And hereof groweth envy, malice, rancour, choler, hatred, displeasure, enmity and what not else; and sometimes fighting, brawling, contention, quarrel picking, murder, homicide and great effusion of blood.'

Having admired the grace and agility of the ice skating Gold Medallists displayed on TV screens, the next time there is a severe winter you may be tempted to take to the ice. Don't do it. For one thing, other skaters will be there to collide with you and gash your legs with their skates. Then small children will make slides, knock you down and fracture your limbs.

Worst of all, the ice will break up, as it did on the lake in Regent's Park in London during what the Victorians called the Great Freeze. On 15th January 1867 some three hundred skaters had — naturally — ignored warnings to keep off the ice and were going through their paces when the ice suddenly detached itself from the land, gave a great heave and shattered into huge chunks. Some of the skaters managed to reach firm ground but the others fell into the lake, which was 12 ft deep in places, desperately clutching lumps of ice to keep themselves afloat. In the words of one eye witness reported in the *Illustrated Times* on 19th January, 'Only those who were on the spot, and saw with their own eyes what took place, can form an adequate idea of the calamity which in an instant placed 200 persons at the very gates of death.' Forty people died that day as a result of going on the ice when warned not to, but nobody was able to explain just why the lake of ice should have disengaged so suddenly.

Archery, now a minority sport, is probably no less dangerous than it was in the days when proficiency with bow and arrow was a civic duty imposed on citizens in a primitive kind of Territorial Army training scheme. A coroner's inquest in July 1508 reported:-

> 'Between 3 and 4 o'clock on 28 June, when John Esterffyld late of Ringmer, labourer, was shooting arrows at targets at Ringmer with

> 4 other honest persons, John Newman was behind one of the targets out of Esterffyld's view and unknown to him. Esterffyld shot and his arrow flew beyond the target and struck Newman on the left side of the head, giving him a wound of which he immediately died. So Esterffyld killed him not maliciously but by mischance and against his will.'

OVERSEAS TRAVEL is an increasingly popular way of spending holidays. In previous years it was almost a fulltime occupation for the leisured classes and in 1789, to help them on their way, Count Leopold Berchtold wrote 'An Essay to Direct and Extend the Enquires of Patriotic Travellers.' He pointed out one of the discomforts that air travellers today will recognise:-

> 'Travellers in carriages are very liable to have their legs swelled; in order to prevent being thus incommoded, it will be advisable to wear shoes rather than boots, to untie the garters, to alight now and then, and to walk as often as opportunity permits, which will favour circulation.'

He also warned against the common but dangerous method of heating bedrooms which we have already mentioned:-

> 'The vapours of charcoal are also exceedingly prejudicial; people should be remarkably careful never to permit a pan of charcoal to be brought into their apartment, unless it is quite burnt to ashes; it would be dangerous to sleep with it in the bed-room, as a great many lives have been lost in that manner.'

These same leisured classes spent a great deal of time at their *toilette* and in the 18th Century it was the fashion for great ladies to wear their hair piled high, powdered and adorned with jewels, flowers, feathers and anything else that fancy might dictate. Contemporary satirists made great fun of the fashion and a print of the day shows a lady being carried in a sedan chair, in the roof of which a hole has been cut to accommodate the towering structure. Even so, the effort required to deal with the following incident, reported in the *Morning Post* on the 9th March 1776, seems excessive:-

> 'About nine o'clock, the head dress of a lady in high life, who lives in the neighbourhood of Portman Square, accidently caught fire, but by the timely assistance of three engines and plenty of water, it was got under a little before twelve.'

A GAME OF CHESS is not considered a perilous occupation but a man did once break his leg in a game. He found the game absorbing and while studying the board wrapped his own leg around the leg of the chair. His opponent then made an unusual and unforeseen move. This surprised him so much that he jumped up immediately and the sudden move broke his leg. He had to go to hospital and lost his game.

Snorkelling too is a gentle way of spending leisure time, with few attendant dangers. It was therefore a stroke of hideous bad luck when a bee settled on the end of a snorkeller's tube just as he breathed in. The bee was sucked down the tube into the man's throat where it stung him. It is hard to think of a more unpleasant way of suffocating.

SOME FIELD SPORTS are notoriously dangerous; hence the stringent regulations and etiquette governing the use of sporting guns. A lucky victim of a shooting accident was Sir James MacDonald. Writing to a friend from the Isle of Skye in August 1761 he says:-

> 'I am shot with a ball in both my feet, yet by the hand of providence the ball was so directed as to break no bone nor any

> considerable sinew. This accident happened this day, sennight when I was shooting a deer. Another gentleman was creeping up behind me when his gun unluckily went off in his hand and gave me one wound on the outward part of the right foot aside the great toe; from hence it passed to the left which it first struck upon the soal under the great toe and then grazed along the hollow of the foot till it made a small wound just beside the ancle, tho' it luckily did not touch the ancle itself.
>
> 'It will be hard for you to conceive how this could be done; but as I was lying flat on the ground if you will lay yourself down on your face and let the right foot be a little behind the left and both of them in a straight line you will see how it happened.'

They don't write accident reports like that any more.

Not so fortunate was a servant who suffered the consequences of the bursting of a gun in 1835. Sir William Knighton, Private Secretary and Privy Purse to George IV, was able to look on the bright side of this accident. He wrote to a friend:-

> 'I am sorry for poor B...'s accident. I do not understand what business he had with William's gun; the guns of gentlemen are generally considered as not to be used without permission or some specific order.
>
> 'This affair, however, I consider to come under the head of a particular providence, for dear William might have used the gun next year, and the most disastrous results might have arisen. God be praised, and make me duly sensible of his great and continual mercies.'

Samuel Pepys heard of a sad little tragedy in May 1663 and noted it down:-

> 'My Lord Hinchingbroke, I am told, hath had a mischance to kill his boy by his birding-piece going off as he was a-fouling. The gun was charged with small shot and hit the boy in the face and about the temples, and he lived four days.'

But most tragic of all gun accidents must be one in July 1793. It was reported at the time:-

> 'On Tuesday died Mr. F. Walsh, of Nottingham, in consequence of a mortal wound he received on His Majestry's birthday from his own son, who sportfully discharged a pistol close to his father; the

wadding unfortunately penetrated his body, beyond the skill of the physician, and he died in inexpressible pain.'

THE *PRINCESS ALICE*, launched in 1865, was a paddle steamer built for pleasure cruising on the Thames, then, as now, a popular activity. For thirteen years she carried holiday makers up and down the river, but the party came to terrifying end on 3rd September 1878. Returning at dusk from an excursion to the Thames estuary with around 750 passengers aboard, she was sliced in two by the *Bywell Castle*, an 890 ton collier steaming down river. The *Princess Alice* sank immediately and despite some heroic rescue attempts about 640 passengers and crew lost their lives.

Eye witness accounts and evidence given at the inquest make blood-chilling reading. The Board of Trade held an inquiry and the owners of the two steamers brought legal actions against each other. There were accusations and counter-accusations, claims and counter-claims. The *Princess Alice* was strongly criticised for carrying too many passengers and insufficient emergency equipment (two lifeboats and twelve lifebuoys for 750 people). But what clearly emerged was the need for more stringent shipping regulations and a strict compliance with them.

The pleasure steamer had been named after Princess Alice of Darmstadt, one of Queen Victoria's daughters. She died of diphtheria on the day the Board of Trade published its report on the disaster.

RIDING A HORSE is a source of great pleasure to many people. So is gambling. These two recreations joined forces many years ago —

formally on the race track, informally among friends. But it can be a dangerous partnership, as one lady found out; her splendidly direct foray into Loss Prevention — in this case the loss of a husband — is related by Captain Gronow in his *Reminiscences and Recollections, 1810–1860*:-

> 'The Bold Wife of a Rash Husband. About thirty years back a bet was made in Paris by the Comte de Chateauvillard that he would ride a horse which no groom would venture to mount because of its vicious propensities. The animal in question had been allowed to remain idle for several months, without having ever been touched by any one during all that time; for it was fed through a hole in a neighbouring stall, and watered and littered in a similar manner.
>
> 'As the time approached for the conditions of the bet to be carried out, great excitement prevailed in the clubs with regard to it, especially among those skilled in horsemanship, and a wager of 20,000 francs was jointly laid by several gentlemen against the Count. Information was, however, conveyed to the Count's wife, an Irish lady by birth, and foreseeing the danger her husband would inevitably incur, she armed herself with a brace of pistols, entered the stable, and placing one of them to the horse's head fired. The animal reared and fell dead, the Lady exclaiming, 'Thank God, I have done my duty!'

There are of course more disreputable ways of spending leisure time than any of those noted above, and we close this chapter by delicately refraining from comment on one of the pitfalls attendant upon such depravity. Let a Coroner's Inquest speak for itself:-

> '20th August 1794. On Saturday evening the coroner's inquest was taken at the Barn public house, St. Martin's-lane, on the body of George Howe, who, on Friday afternoon, threw himself from a three-pair-of-stairs window in Johnson's court, Charing-cross, and was killed on the spot. The jury returned their verdict, 'Accidental death in endeavouring to escape from illegal confinement in a house of ill fame'.'

# CHAPTER 4

# ON THE MOVE

IT CAN SAFELY BE ASSUMED that travel and transport have always presented hazards, and with great confidence we can predict that new means of travel — whether by land, sea or air — will generate fresh safety problems.

Even walking can be dangerous, as a Coroner's inquest noted in 1554:-

'Between 5 and 6 on 27 February, when Richard Jeffery of Chiddingly was walking in his backyard in Chiddingly, by misadventure he fell in the mud and, returning home with his face defiled with mud and mud obstructing his eyes, by misadventure he fell into a pond and was drowned.'

SINCE MAN FIRST TAMED THE HORSE he has used it for travel, along with oxen, camels and the like. But a horse needed the same careful maintenance as a space shuttle, and as Benjamin Franklin pointed out in the 18th century: 'A little neglect may breed mischief... for want of a nail the shoe was lost; for want of a shoe the horse was lost; and for want of a horse the rider was lost.'

One rider who was lost, we were taught at school, was our own William III, he of the Glorious Revolution. In 1702 his mount stumbled on a molehill and threw its rider, breaking the royal collarbone. Death followed soon after; hence, of course, the Jacobite toast to 'the little gentleman in the velvet coat.'

John Aubrey, the enthusiastic 17th century biographer, seems to have been accident-prone when astride a horse, as the following passages from his own autobiography attest:-

'On Monday after Easter weeke, my uncle's nag ranne away with me, and gave me a very dangerous fall... Then (I thinke) June 14, I had a fall at Epsam, and brake one of my ribbes and was afrayd it might cause an Apostumation.'

*and a year or two later:-*

'March or April, like to break my neck in Ely Minster, and the next

> day, riding a gallop there, my horse tumbled over and over, and yet (I thanke God) no hurt.'

and, more worrying:-

> 'Munday after Christmas, was in danger to be spoiled by my horse, and the same day received laesio in Testiculo which was like to have been fatall.'

WITH TRAFFIC ACCIDENT STATISTICS BEFORE US, is it uncharitable to hope that the man who first invented the wheel fell under one and was crushed to death? Such a fate has been common over the centuries, and in writing to Lord Herbert in September 1770 the Earl of Pembroke was able to relate with no apparent expressions of horror:-

> 'In coming this morning from Maiden Early, Baron Eyre's, I found in the middle of the Turnpike, a dead boy, not yet cold, with his head crushed literally as flat as a flounder. He slipt, it seems, from the shafts of the waggon which he was sitting, under a broad wheel, which went over him. I hear he is a Salisbury boy.'

Hit-and-run accidents did not come in with the motor car. At a Coroner's Court in 1337 it was stated:-

> 'On Thursday, about the hour of vespers, two carters taking two empty carts out of the city were urging their horses apace, when the wheels of one of the carts collapsed opposite the rent of the hospital of St. Mary, Bishopsgate, so that the cart fell on Agnes de Cicestre, who immediately died. The carter thereupon left his cart and three horses and took flight in fear, although he was not suspected of malicious intent.'

Nor is suing for damages a 20th century preoccupation, and the compensation awarded to this poor lady coach traveller seems paltry indeed by today's standards. On 22nd February 1785 an action was brought by a Mr. Rowley against Mr. Sabin, the proprietor of the Croydon stage, on which Mr. and Mrs. Rowley had been passengers some five months previously. It seems the coach was overloaded. '... the coachman, instead of stopping at the Swan at Charing-cross, drove past it, and, in endeavouring to turn round, overturned the coach, by which the plaintiff's foot was so much bruised that it was obliged to be taken off, and she has been ever since confined in the hospital'. A verdict was given for the plaintiff, with £100 damages and costs. Lord Loughborough, before whom the case was heard in the Court of Common Pleas, remarked:-

> 'If an accident happened by the overloading the coach, the master of such coach will be liable to the damages, he overloading the coach for his own emolument.'

A dictum as applicable today as it was then.

AN EXOTIC INCIDENT was reported on 20th October 1816. The mail coach from Exeter to London had just left Salisbury when what appeared to be a calf was seen to be loping along beside the horses and causing them some distress. As the coach pulled up at the Pheasant Inn, between Stockbridge and Salisbury, the guard on the coach hauled out his blunderbuss to shoot the creature. It then became apparent that it was a lioness. It attacked one of the horses, called 'Pomegranate', by putting its claws around the horse's neck and going for the throat. The other horses lashed out at the lioness and might have driven her off had not a group of men arrived with a mastiff which they set on the lioness. The dog was no match for the lioness, who quickly killed it and dragged the body away to a nearby granary where the lioness was recaptured.

Pomegranate survived the attack and became a fairground exhibit, presumably at the same fairground from which the lioness was unleashed.

A passenger who had been brushed by the lioness 'went off his head' and was confined to an asylum for the next twenty-seven years. This bizarre event was commemorated by a stamp issued by the Royal Mail in 1984 and should be noted by those carrying out risk assessments on transportation problems.

DRINK-DRIVING caused a mail coach accident on the Threcastle to Llandovery road on 19th December 1835 and a pillar has been erected in memory of the event. The inscription states:-

> 'This Pillar is called Mail Coach Pillar and erected as a caution to mail coach drivers to keep from intoxication, and in memory of the Gloucester and Carmarthen mail coach which was driven by Edward Jenkins on the 19th day of December in the year 1835, who was intoxicated at the time and drove the mail on the wrong side of the road, and going full speed or gallop met a cart and permitted the leader to turn short round to the right hand and went down over the precipice 121 feet, where, at the bottom near the river, it came against an ash tree when the coach was smashed into several pieces. Col. Gwynne of Glan Brian Park, Daniel Jones Esq. of Penyboat, and a person of the name of Edwards were outside and David Lloyd Harries Esq. of Llandovery solicitor and a lad of the name of Kernick were inside passengers by the mail at the time, and John Compton, Guard.'

A coach was involved in a double disaster at Manchester in 1798. As reported on 26th November of that year:-

> 'A coachdriver, late last night (near the hour of twelve), drove his vehicle into our river, near the Old Bridge, for the purpose of washing; when, the current running strong, the horses were soon driven into the centre of the stream, forced under one of the arches, and in that state (too shocking almost to conceive) they swam with the man on the box, through Blackfriars-bridge, fighting and struggling for their lives till one in the morning. The poor fellow, in his endeavours, had entangled his legs in the reins; but from them he extricated himself with a knife; when, fortunately coming nearly in contact with a dyer's flat, he, by an astonishing effort, jumped from the box on the same, where he lay several minutes in a state of insensibility.
>
> 'The horses, after swimming about the river some time, followed

their master to the flat, and attempted to raise their fore-feet upon it: the poor man, with the little strength he had left, held up the head of one of the creatures, till, with a convulsive groan, it expired in his arms. From the active assistance of several persons, attracted by the cries of the coachman, they had so far succeeded in securing the other horse as to extricate him from the reins, and had got him nearly half up Mrs. Duxbury's steps, when, owing to the tempestuousness of the night, he slipped from their holds, and again plunged into the river; after which nothing more was seen of him.

'Happy would it have been had the calamity ended here: curiosity (early in the morning following) called crowds of people together to see the bodies of the horses floating; among others a group of nine or ten women and children very incautiously got together on a dyer's stage, hanging over the river near the New Bridge; when, shocking to relate, the bottom of the stage gave way, and they were all in an instant precipitated into the river. Three were recovered before life was gone; the strength of the current rendered every endeavour to save the others ineffectual, and they were all swept away! On how slender a thread does human life hang! the insecurity of these stages, from the number of years they have been erected, renders it a matter of astonishment that even an individual will trust his person thereon.'

A COACH AND HORSES is not normally encountered on or in a river, nor are ice floes — in London at all events†. But in January 1795 it was reported that lumps of ice brought up the Thames severed the cables by which two vessels were moored just below London Bridge. The tide drove them against the bridge with such force that one of the ships, a three-master, hit the centre arch, lost her masts, knocked down two of the lamps on the bridge, continued through the arch 'with incredible velocity' and on through Blackfriars Bridge to Somerset Place, where she ran aground. The crew, sensibly, had taken to the boats before the vessel reached London Bridge.

And one year later, in 1796, another foreign body presented itself as a shipping hazard on the Thames. In August of that year a young 19 ft whale reached Rotherhithe, where it was killed after overturning two vessels.

† The Chronicle of Dunstable Priory narrated a hard frost in 1281 and 'near Biddenham bridge the ice broke under a certain woman and she was stranded on a floe; and from there she was carried resting firmly on the water as far as Bedford bridge: nor could anyone come to her aid. At Bedford bridge the ice broke and the woman was not seen again.'

Those were exciting times for shipping on the Thames and an accident with altogether jollier results for marine life was reported on 17th August 1794:

> 'Yesterday morning about two o'clock, a fire broke out on board the Neptune, West-India ship, lying in the Pool. Her cargo, no part of which had been landed, consisted entirely of rum. She was immediately towed out of the tier, and run ashore on the Southward side. She burnt very furiously till late in the evening, but without extending the calamity to other vessels. By the quantity of rum destroyed on board the Neptune, the fish in the Thames were so affected as to float up with the tide in such numbers that they were collected by the people, on both shores, in baskets full.'

THE HAZARDS OF OVERSEAS TRAVEL in the days of sail are too numerous to dwell upon. A typical disaster took place in 1788, when a ship carrying 113 passengers and crew sprang a leak *en route* from Bengal to Madras. 'Whilst the vessel could be kept clear by the pumps, no danger was apprehended; in the evening however, the pumps were rendered useless from being choaked with rice, of which the cargo consisted; from this moment nothing but the most dismal prospect presented itself.' Only 48 persons survived.

Not so typical was a danger that faced one ship's captain in the same year:-

> 'Friday last arrived at Lynn in Norfolk, the Archangel, from Greenland, Captain Cook... When this ship was in Greenland, Captain Cook, the surgeon and mate, went on shore, when the captain was seized by a monstrous bear, which immediately hugged him with his paws; the captain called to the surgeon to fire at the creature, though at fifty yards distance, which he did, and fortunately shot the bear through the head, which instantly killed it, and Captain Cook was by this means saved from being torn in pieces.'

With all the sinkings at sea and the loss of cargoes valuable to a trading nation, salvage operations had to be attempted. They had their own perils, as the following report dated 1st June 1783 indicates:-

> 'The ingenious Mr. Spalding, accompanied by one of his young men, went down twice in his diving-bell at the Kish bank, Ireland, where the East Indiaman was sometime since wrecked, for the

purpose of recovering some of her materials. He did nothing more, however, than examine her situation, etc., determining to go to work next morning. Accordingly, Monday morning, about six o'clock, he and his young man went down, and continued under water about an hour, in which two barrels of air had been sent down for the supply of the bell; but a deal of time having elapsed without any signal from below, the people on deck, apprehensive that all was not right, drew up the bell, and Mr. Spalding and his young man were both discovered to be dead.'

SAIL WAS SUCCEEDED BY STEAM, but doubts about safety were being expressed. Thomas Raikes, from a rich and respected family of City merchants, wrote in his diary on 12th June 1837:-

'An accident has lately happened in Hull, which may prove the advantages of steam navigation are not unattended by proportionate risk and danger. The steam vessel *The Union* was the scene of this catastrophe. From some negligence in heating the apparatus, the boiler burst, and out of 120 passengers, not more than twelve or fifteen were saved from death. The vessel went to the bottom.'

But things were not quite as black as Raikes painted them and the great engineer Isambard Kingdom Brunel built the *Great Western*, the first steam powered ship designed for the North Atlantic run. There were some initial problems. After fitting out on the Thames in 1838, with Brunel abroard, she steamed for Bristol to take on passengers and supplies before setting out for North America. Just off the Nore fire broke out; poor boiler lagging† beneath the base of the funnel caused the deck beams and the underside of the deck planks to catch fire. Brunel, determined to help fight the fire, raced to the ladder to get below. A partly burnt rung gave way and he fell 18 ft, his fall happily being broken when he landed on the Managing Director of the Company which owned the ship who was playing a hose on the fire at the foot of the ladder. Brunel, seriously injured, was put ashore at Canvey Island. The fire was extinguished and the *Great Western* continued to Bristol. She left there in April 1838 and reached New York some fifteen days later, not quite the first steamer to cross the Atlantic; she was pipped to the post by the *Sirius*.

†Poor indeed — the lagging had been treated with a highly combustible mixture of oil and red lead.

THE INTERNATIONAL MARITIME ORGANISATION (IMO) is concerned with regulation of international shipping, under the auspices of the United Nations, and what follows, tragic though its consequences were, must raise a smile.

The incident took place on 6th December 1917 at Halifax, Nova Scotia. The French ammunition ship *Mont Blanc*, inward bound for the Bedford basin to join a convoy crossing the Atlantic, carried some 5 million lbs of explosives. In addition, nearly half a million lbs of benzene in drums was stowed on deck.

She collided with a Belgian ship and immediately caught fire from the spilt benzene. The crew quickly abandoned ship and rowed for the shore, while H.M.S. *Highflyer* sent two boats to the aid of the *Mont Blanc*. Seventeen minutes after the collision and fire there was a massive explosion. Parts of the *Mont Blanc* were scattered over a wide area of Halifax and the suburb of Richmond. The aft gun landed nearly two miles away and part of the anchor was found four miles away. Buildings within a radius of one and a half miles suffered severe structural damage, and ninety-five per cent of all glass in the city was broken. A tidal wave engulfed two blocks of the city harbour and there was no trace of the two boats from H.M.S. *Highflyer*; panes of glass 60 miles away were shattered, and the explosion was heard 190 miles away. Eight thousand people were injured and there were 1,800 fatalities.

And the name of the Belgian ship that collided with the *Mont Blanc*? She was called the *Imo*.

SOONER OR LATER travelling underneath the sea would become a practical proposition. The world's first mechanically driven submarine was, inappropriately as it turned out, named *Resurgam* (I Shall Arise). Built at Birkenhead in 1879, she was under tow to Portsmouth for Admiralty trials when a violent storm off the North Wales coast sent her inventor and two-man crew scurrying by boat to the yacht towing her. The submarine's conning tower could not be sealed from the outside; the heavy seas broke over the tower and swept down the hatch, increasing the weight until the tow hawser broke. *Resurgam* sank irretrievably to the bottom. But this was not the end of the story. The yacht was forced to take shelter up the River Dee, where she dropped anchor. Another gale blew up, the yacht parted her chains and went adrift. She signalled for help. A passing ship came to the rescue but managed to ram the yacht and turn her also into a total wreck.

THE INTRODUCTION OF THE RAILWAY SYSTEM, with its steam engines, opened a new chapter of accidents. Being run over by a train and a boiler explosion were the two most serious. Indeed, at the opening of the first railway system in the U.K. in 1830, the Stockton to Darlington railway, the Minister for the Board of Trade was killed by Stephenson's *Rocket*.

For an eye-witness account of this celebrated tragedy we can turn to that indefatigable gossip, diarist and letter writer, Thomas Creevey. He wrote to his stepdaughter on 19th September 1830:-

> '... Jack Calcraft has been at the opening of the Liverpool railroad and was an eye-witness of Huskisson's horrible death. About nine or ten of the passengers in the Duke's car had got out to look about them, whilst the car stopped. Calcraft was one, Huskisson another, Esterhazy, Billy Holmes, Birch and others. When the other locomotive was seen coming up to pass them, there was a general shout from those within the Duke's car to those without it, to get in. Both Holmes and Birch were unable to get up in time, but they stuck fast to its sides, and the other engine did not touch them. Esterhazy, being light, was pulled in by force. Huskisson was feeble in his legs, and appears to have lost his head, as he did his life. Calcraft tells me that Huskisson's long confinement in St. George's Chapel at the King's funeral brought on a complaint that Taylor is so afraid of, and that made some severe surgical operation necessary, the effect of which had been, according to what he had told Calcraft, to paralyse, as it were, one leg and thigh. This, no doubt, must have increased, if it did not create, his danger and caused him to lose his life. He had written to say his health would not let him come, and his arrival was unexpected. Calcraft saw the meeting between him and the Duke [Duke of Wellington] and saw them shake hands a very short time before Huskisson's death. The latter event must be followed by important political consequences...'

Another contemporary commented on William Huskisson's death: 'There were perhaps 500,000 people present on this occasion, and probably not a soul besides hurt... It is the more remarkable because these great people are generally taken such care of, and put out of the chance of accidents.' So great an impact did this accident have on public imagination (not to mention the political fallout from the loss of a potential Prime Minister) that it was being celebrated in verse a quarter of a century later (see chapter 14).

A railway accident from the same period is commemorated on a tombstone in the graveyard of St. Mary's Church at Harrow-on-the-Hill, and is a reminder to us of what lay in store for the victim of an accident in those days.

*To the Memory of*
*THOMAS PORT*
*Son of John Port of Burton upon Trent*
*In the County of Stafford, Hat Manufacturer,*
*Who near this town had both his legs*
*Severed from his body by the Railway Train.*
*With the greatest fortitude he bore a*
*Second amputation by the surgeons and*
*Died from loss of blood,*
*August 7th 1838, Aged 33 years.*

*Bright rose the morn and vig'rous rose poor Port*
*Gay on the Train he used his wonted sport:*
*Ere noon arrived his mangled form they bore,*
*With pain distorted and o'erwhelmed with gore:*
*When evening came to close the fatal day,*
*A mutilated corpse the sufferer lay.*

EXPLOSIONS IN EARLY TRAIN ENGINES were common (as indeed they were in industrial steam boilers), and the Railway Inspectorate was set up to investigate such accidents.

The first report they received concerned an accident which occurred at Bromsgrove on 10th November 1840 on the Birmingham and Gloucester Railway. The driver and foreman were killed in the explosion, which was attributed to the 'boiler plate not being strong enough'. The engine was named *Surprise.*

Those killed in that explosion have headstones with the following inscriptions:-

*To the memory of Joseph Rutherford*
*who died November 11th 1840, Aged 32 years.*
*My time was spent like day in sun*
*Beyond all cure my glass is run.*

and

To the Memory of Thomas Scaife, late an engineer on the
Birmingham and Gloucester Railway who lost his life at

Bromsgrove Station by the explosion of an Engine Boiler, Tuesday the 10th November 1840. He was 28 years of age, highly esteemed by his fellow workmen for his amiable qualities and his death will be long lamented by all those who had the pleasure of his aquaintance. The following lines were composed by an unknown Friend as a Memento of the worthiness of the Deceased.

*My* engine *now is cold and still,*
*No water does my* boiler *fill;*
*My* coke *affords its flame no more,*
*My days of usefulness are o'er.*
*My* wheels *deny their wonted speed,*
*No more my guiding hands they heed;*
*My* whistle, *too, has lost its tune,*
*Its shrill and thrilling sounds are gone;*
*My* valves *are now thrown open wide,*
*My* flanges *all refuse to guide,*
*My* clacks, *also, tho' once so strong,*
*Refuse to aid the busy throng.*
*No more I feel each urging breath,*
*My* steam *is now condensed in death.*
*Life's* railway *o'er each station past,*
*In death I'm stopped and rest at last.*
*Farewell, dear friends, and cease to weep,*
*In Christ I'm safe, in Him I sleep.*

There was a doubly tragic accident on a North American railway in September 1906. A freight car was standing at the railroad sidings of Jellico, Tennessee, with a load of 150 cases of 60% dynamite and 300 cases of 40% dynamite, a total of 22,500 lbs of explosives. The freight car was rammed by two others containing pig iron, with such force that the dynamite exploded. Eleven people were killed and a crater 30 ft deep by 75ft diameter was left.

A man called Lee Hill was one of those killed in the explosion. His father, sister, wife and four children came to Jellico to claim his body and to return with it for burial at their home town in South Carolina. They boarded the train but it collided with another one at New Market in Tennessee and all seven were killed. They are buried in Gaffney in South Carolina.

**THE IDEA OF FLYING LIKE A BIRD** caught man's imagination at a very early date. A visionary who sought to go one better than Icarus was an 11th century monk, Brother Eilmer. He was a mathematician and scientist, and made lengthy plans for his flight. He eventually descended from one of the towers of Malmesbury Abbey with wings strapped to his arms and legs, but crashed and broke both legs. According to Milton, he was 'so conceited of his art, that he attributed the cause of his fall to the want of a tail, as birds have, which he forgot to make to his hinder parts.' Nevertheless he survived to a great, if lame, old age.

Monastic orders also took to the air in Scotland, where they were fortunate to have a king who encouraged new thinking in arts and sciences. During the reign of James IV (1488–1513) Abbot Damien of Tongland made a pair of wings from hens' feathers and took off from the walls of Stirling Castle. He landed in a midden. However, he suffered only a broken leg and blamed his failure to soar on not having used eagles' feathers for his wings.

That early version of hang gliding had a happy outcome, and broken legs seem to have been the only ill consequences of a later flying accident. In one of his *Brief Lives*, John Aubrey records:-

> 'I remember Sir Jonas told us that a Jesuite (I think 'twas Grenbergerus, of the Roman college) found out a way of Flying, and that he made a youth performe it. Mr. Gascoigne taught an Irish boy that way, and he flew over a River in Lancashire (or thereabout) but when he was up in the ayre, the people gave a shoute, whereat the boy being frighted, he fell down on the other side of the river, and broke his legges, and when he came to himselfe, he sayd that he thought the people had seen some strange apparition, which fancy amazed him. This was anno 1635, and he spake it in the Royall Societie, upon the account of the Flyeing at Paris, two yeares since.'

BUT a still later attempt to conquer the air had a more unfortunate outcome. The *New Annual Register* for 1786 described the misadventure:-

> 'Newcastle upon Tyne September 20.
> Lunardi's attempt to ascend yesterday from the Spital ground was productive of a very melancholy accident. The balloon was about one-third full, and a great many gentlemen were holding it by the netting, when Lunardi went to pour into the cistern the rest of the oil of vitriol destined for the purpose. This having caused a strong effervescence, generated inflammable air with much rapidity, that some of it escaped from two different parts of the lower end of the apparatus, and spread among the feet of several gentlemen who

were holding the balloon, and who were so alarmed, that leaving it at liberty, they ran from the spot. The balloon now rose with great velocity, carrying up with it Mr. Ralph Heron, a gentleman of this town, about twenty-two years of age, son of Mr. Heron, under-sheriff of Northumberland.

'This unhappy victim held a strong rope which was fastened to the crown of the balloon, twisted about his hand, and could not disengage himself when the other gentlemen fled; he was of course elevated about the height of St. Paul's cupola, when the balloon turned downward, the crown divided from it, and the unfortunate gentleman fell to the ground.

'He did not expire immediately, having fallen upon very soft ground; he spoke for some time to his unhappy parents, and to the surgeons who came to assist him, but his internal vessels being broken, he died about an hour and a half after the fall.

'Lunardi made a precipitate retreat from the town to avoid the resentment of the populace.'

A couple of years before this Vincent Lunardi himself had written *An Account of the First Aerial Voyage in England*. He mentioned a little difficulty over in-flight meals:-

'When the thermometer had fallen from 68° to 61°, I perceived a great difference in the temperature of the air. I became very cold and found it necessary to take a few glasses of wine. I likewise ate the leg of a chicken, but my bread and other provisions had been rendered useless, by being mixed with the sand, which I carried as ballast.'

Ballooning was very much in vogue in the 18th century, but associated activities such as parachuting also drew large crowds. In March 1790 a certain Mr. Murray, described by the press as 'an eminent optician and man of science', descended in a parachute from a church tower in Portsmouth and 'came to the ground without receiving the smallest injury'.

His repeat performance a month later was not so successful. It was reported on 20th April:-

'Mr. Murray, who some time ago descended from Portsmouth church tower, in a parachute, on Wednesday came down from the Bell Tower of Chichester Cathedral, but not with the same success. When about 14 feet from the top, a sudden gust of wind laid this bold adventurer and the apparatus in a horizontal position; when on a level with the gutter of the cathedral he righted, but an eddy wind threw him a second time horizontally, in which situation he fell to the ground with great force. The blood gushed from his ears, nose and mouth, very plentifully; and he was carried to the Blue Anchor Inn without any signs of life. Four gentlemen of the faculty instantly went to his assistance; and in something more than four hours, animation returned, and in six his speech was restored. Mr. Prescot, one of the above gentlemen, on Friday pronounced him out of danger.'

THE FOLLOWING CENTURY saw much ingenious effort put into the design and construction of steerable balloons, or dirigibles as the early airships were called.

As early as 1852 a Frenchman, Henri Giffard, built a steam-powered airship which incorporated such safety features as a good distance between the flammable envelope and the gondola carrying the engine and pilot, a boiler fire door screened by a gauze like a miner's lamp, and a chimney pointing downwards and rearwards to deflect sparks away from the envelope. The airship made a safe landing after a flight of seventeen miles — not a great distance, but it was man's first flight in a powered machine.

But steam engines with their boilers were too heavy. Gas powered engines were useless. The electrical batteries available at the time were heavier than boilers. Then, at the tail end of the 19th century, the internal combustion engine was invented, for good or evil.

In June 1897 a much heralded and Kaiser-encouraged German airship, powered by one of Herr Gottlieb Daimler's new engines, made a catastrophic demonstration flight in Berlin before assembled members of the armed services and diplomatic corps. As things turned out, all it demonstrated was the folly of disregarding safety standards already established by the Giffards of the aeronautical fraternity. The airship was manned by her designer and a mechanic. On release she climbed rapidly to 3,000 ft, seemingly out of control, and shortly afterwards flames were seen to shoot from the engine and ignite the envelope. The engine exploded and the whole airship came down in flames; the charred bodies of the two aeronauts were found in the wreckage.

ONCE THE AEROPLANE as we know it today had taken to the air, a new accident list had to be opened. Top of the list, as the first passenger casualty, was Lt. Thomas C. Selfridge who was killed on the 17th September

1908 when Orville Wright's aircraft crashed because of a splintered propellor. Wright suffered only a broken leg.

Another early victim of this new means of travel is remembered in the Parish Register of Bowerchalke in Wiltshire:-

> Died April 25 1917. Fredderick Charles Butler 2nd Lt of Rookhay Farm, Bowerchalke, aged nearly 20 years. Killed by a fall from an aeroplane at Dover. Buried April 29. 1917 at 2.30 pm.

In the early days of flying, when cockpits were open to the four winds, most pilots preferred not to use any kind of safety belt. They took the view that it was better to be thrown clear of a crashing aircraft than to be trapped inside the wreckage. What would they have made of the crash-proof aircraft built by yet another flying Frenchman? It was not all that crash-proof, as *The Times* reported on the 15th August 1932 under the headline 'INVENTOR PUSHED OVER A CLIFF':-

> 'M. Albert Sauvant, the inventor of what he claims to be a 'crash-proof' aeroplane, yesterday continued his series of personal demonstrations of the machine by being hurled over an 80 ft cliff at Nice. He was badly shaken, and it was at first thought that he had broken an arm, but medical examination showed that he had suffered nothing more serious than severe bruising.
>
> 'The experiment closely resembled that carried out in March, when M. Sauvant made a vertical drop of 70 ft and emerged unscathed. Yesterday some friends obligingly pushed the machine with him in it over the cliff. After the crash M. Sauvant's friends peered over the ledge, and waited confidently for his triumphant emergence from the wreckage. When it became evident that he was in trouble several men descended by ropes to his rescue. With much difficulty they managed to drag him clear. Later, when he had recovered, M. Sauvant declared that he was delighted with the success of his experiment since he would certainly have been killed in an ordinary machine.'

Let the press have the final say on what can happen to the travelling public. The *West Wales Guardian* reported an incident under the informative headline:-

> 'Bus on Fire — Passengers Alight'

# CHAPTER 5

# GUNPOWDER PLOTS AND OTHER EXPLOSIONS

THE OXFORD ENGLISH DICTIONARY DEFINES *EXPLOSION* — in the sense we are concerned with — as follows:-

'Of a gas, gunpowder, etc. The action of 'going off' with a loud noise under the influence of suddenly developed internal energy.'

This is a broad definition (it could be applied to the rebellious teenager storming out of home, ghetto blaster in hand), but explosions as we know them have been the occasion of death and injury for centuries. They range in intensity from the common firework to the mushroom-shaped cloud.

ON A SCALE OF ONE TO TEN, the following grotesque incident might come pretty low down, but then we are unaware of the fate of the patient. It seems that when a surgeon was sealing a small incision in the lower digestive tract with an electrocautery instrument, he managed to ignite the patient's intestinal gas. The explosion threw the doctor across the operating theatre.

Higher up our scale, if we are to believe 18th century reports, the earth was bursting out all over. At Tipperary in Ireland, in March 1788, a bog of some 1,500 acres 'began to be agitated in an extraordinary manner, to the astonishment and terror of the neighbouring inhabitants.' A few days later it burst and a 'kind of lava' spewed out. Vast tracts of land were laid waste and four houses totally destroyed. And in July of the same year, at Cotencin on the north west coast of France, a hill of considerable size 'suddenly burst with an explosion that shook the adjacent parts two miles round, and immediately a torrent of water mixed with sand, earth and broken stones issued from the opening, and inundated the fields to the depth of several feet, sweeping away cattle, farm-houses and cottages on its course to the sea.'

In 1797 there was another explosion of the same nature, puzzling at the time. As reported in the *New Annual Register* on 21st January of that year:-

'This night at 11 o'clock, a cottage at Newtown Ferrers, about eleven miles from Plymouth, in which slept an industrious widow (cottager) and her two children, was overwhelmed by the bursting of a very large field and orchard on a hill above the cottage, in Memblard Lane. It totally destroyed the cottage and a barn, and suffocated the widow and her two children, who were found dead under a very great heap of earth, elm trees, and cider trees. A large chasm in the field above the cottage was found, out of which issued a rivulet of water. The farmers imagine it was owing to the bursting of a spring that this accident happened. The bodies were dug out on Monday; and Mr. Whitford, coroner for the southern district of Devon, took an inquisition, and the jury returned a verdict, 'Accidental death'.'

THE FACTORY WORKER who reported the following improbable accident had no doubts as to its cause. He worked in a nitroglycerine factory, and alleged that a batman in the Prussian army had discovered a new use for the explosive. If wiped over black boots it made the leather supple and gave a good shine without hard buffing. Unfortunately, when the batman's booted officer saluted a superior he clicked his heels hard and in the explosion that followed lost both legs.

Compare our next nitroglycerine incident. In 1886 Mr. Wilson P. Foss was manager of the Clinton Dyanamite Company's plant at Plattsburgh, New York. He was inside a building and standing four yards from a wash tank that contained 900 lbs of nitroglycerine when an accidental gush of live steam from an open valve detonated the nitroglycerine, destroying the building and leaving a crater 30 ft deep. Workers at the plant hurried to the site, expecting to find Mr. Foss's remains among the wreckage. He had, however, been blown by the force of the explosion out of sight round the bend of a nearby frozen river, and to the startled incredulity of the workforce he reappeared to them striding on the ice around the river bend. The truly astonishing feature of this accident is that, whereas the Prussian officer lost both his legs from just a smear of nitroglycerine, all that Mr. Foss was deprived of by 900 lbs of the stuff was his clothing below the waistline. His worst injuries seem to have been the hundreds of imbedded splinters from the spruce trees he landed in.

Before we get down to the serious business of gunpowder a note of what else blows up overseas may not be out of place. It was recently reported that a Romanian farmer was injured when a pig he had slaughtered for Christmas exploded. The farmer, from Cluj, in

Transylvania, had inflated the pig with butane gas to make it easier to clean the skin.

Whether or not gunpowder was invented by the Chinese, it has been the cause of much loss of life, even when used for peaceful purposes. This entry from the Parish Register of Breage probably refers to its use in quarrying operations:-

> 'Thomas Epsly of Chilcumpton parish, Sumersitsheere. He was a man who brought that rare invention of shooting the rocks which came heare in June, 1689, and he died at the bal [i.e. mine] and was buried at the breag on the 16 day of December 1689.'

and Samuel Pepys' servant does not seem to have had any martial thought in mind on 2nd November 1661, the diary entry for which day notes:-

> 'This night my boy Wainman, as I was in my chamber, overheard him let off some Gunpouder; and hearing my wife chide him below for it, and a noise made, I call him up and find that it was powder that he had put in his pocket, and a match carelessely with it, thinking that it was out; and so the match did give fire to the powder and had burned his side and his hand, that he put into his pocket to put out the fire. But upon examination, and finding him in a lie about the time and place that he bought it, I did extremely beat him. And though it did trouble me to do it, yet I thought it necessary to do it. So to write by the post, and to bed.'

But the manufacture and storage of gunpowder were known to be dangerous undertakings in the centuries when it was a common weapon of war. In 1650, in *The Great Art of Artillery*, Casimir Simienowicz warned that gunpowder should be handled with a 'continual eye upon Heaven ... For the accidental Shock of two Stones, the hasty Attrition of two Strings, nay the very impetuous Rubbing together of two Straws, may be the Death of you ...'

A warning to be heeded. In 1878, in a lead mine, one Thomas Harrison was working on a rich vein. He had drilled holes to take the gunpowder which would blast the rock. He had inserted the powder and pushed in the long iron needles or 'prickers' to form holes for the fuses, packing them round with shale and clay until each hole was full. He then began to withdraw a pricker, but the iron needle struck some bare rock and caused a spark which ignited the gunpowder. In the explosion Harrison was fatally wounded — he was just nineteen years old.

AN EARLY AND DISASTROUS EXPLOSION occurred at the gunpowder plant at Delft in the Netherlands. At half past ten on Monday, 12th October 1654, some 85,000 pounds of gunpowder were ignited. The ensuing five explosions completely destroyed more than 500 houses, blasted a crater 15 ft deep, and were heard on the Isle of Texel, a hundred miles away. One hundred people were killed, of whom only 53 were identified. Many children died when two schools were destroyed. Ten days later the remains of the dead were still being gathered. Survivors were being rescued up to four days after the explosion, and remarkably a year old girl was found alive, still in her little chair, with walls and ceiling tottering precariously just over her head. Some 1,000 people were injured, and doctors from the Hague and Rotterdam came to assist. The exact cause of the accident is unknown, but it was established that a civil servant had gone to the plant a quarter of an hour before the explosion to get a sample of the gunpowder. It was suspected that sparks from the big iron locks had ignited the gunpowder.

THE QUEEN OF BOHEMIA (sister to our Charles I) mentioned this explosion in a letter she wrote from the Hague on 19th October 1654 to Sir Edward Nicholas, Secretary of State:-

'... I was at Delft to see the wrack that was made by the blowing up of the powder this day sevenight, it is a sad sight, whole streets quite razed; not one stone upon another, it is not yett knowen how manie persons are lost, there is scarse anie house in the toune but the tyles are off ...'

A hundred years later, on 26th July 1757, Horace Walpole also wrote to a friend:-

'We have given ourselves for a day or two the air of an earthquake, but it proved an explosion of the powder-mills at Epsom. I asked Louis [a servant] if it had done any mischief: he said, 'Only blown a man's head off:' as if that was a part one could spare!'

On 7th January 1772 Mr. Walpole had another explosion to write about. A less laconic letter to General the Hon. H. S. Conway, Lieutenant General of the Ordnance, was also written from his beloved Strawberry Hill, within a few miles of the Hounslow powder mill:-

'You have read of my calamity without knowing it, and will pity me when you do. I have been blown up; my castle is blown up; Guy Fawkes has been about my house; and the fifth of November has fallen on the 6th of January! In short, nine thousand powder-mills broke loose yesterday morning on Hounslow-heath;... As lieutenant-general of the ordnance, I must beseech you to give strict orders that no more powder-mills may blow up. My aunt, Mrs. Kerwood, reading one day in the papers that a distiller's had been burnt by the head of the still flying off, said she wondered they did not make an act of Parliament against the heads of stills flying off. Now, I hold it much easier for you to do a body this service; and would recommend to your consideration, whether it would not be prudent to have all magazines of powder kept under water till they are wanted for service. In the meantime, I expect a pension to make me amends for what I have suffered under the government.'

It is not to be supposed that Horace Walpole received any compensation, nor were the Government able to ban gunpowder explosions. They continued with relentless regularity, and the following report, dated 14th October 1790 is typical of the period:-

> 'On Tuesday afternoon at four o'clock, the inhabitants of Dartford and of the country for several miles around, were alarmed by a dreadful concussion, occasioned by the explosion of 70 barrels of gunpowder at the works of Messrs. Pigon and Andrews, situated about a short mile to the southward of Dartford.
>
> 'No certain account can be given of the manner the fatal spark was communicated to the gunpowder in the corning-house, which was the first building that blew up. No work was carrying on in that place at the time in which the accident happened; it is therefore supposed that some electric fire had entered the building and ignited the loose gunpowder. From hence the explosion instantly communicated itself to the stoves, to five powder mills, and to a close magazine containing 25 barrels of gunpowder, which, from their confined state, spread dreadful devastation around...
>
> 'Six men were destroyed in the dreadful havoc, most of whom have left wives and families behind them. The foreman of the works has left a wife and seven children.
>
> 'It is remarkable that the runners, axletrees, and the wheels of the mills, have received very little damage, so that it is thought some of them will be in a condition to be worked in about a fortnight.'

The factory was soon back in production, and soon blowing up again. On 14th January the following year there was another explosion. As reported:-

> 'The corning-mill belonging to the gunpowder works of messrs. Pigon and Co. at Dartford, this day blew up, by which unfortunate accident two men and a boy were killed. ... One man had fortunately left the mill not more than a minute before the explosion took place; and what, though singular, is true, this is the third time he has thus miraculously escaped from similar accidents.'

In April 1798 the gunpowder mills at Battle also blew up and seven separate buildings were totally destroyed. The report on this explosion concluded with a mention of precautions to be taken when rebuilding the factory:-

> 'A number of workmen are at present employed in clearing the ruins, in order to erect new buildings, which we understand are to be at secure distances from each other and in other respects so contrived as to be rendered less liable to communicate fire from one to the other in case of an accident in either.'

Evidently a hazard had been identified.

ANOTHER HAZARD that has been identified is, of course, tobacco smoking but we doubt whether the Euroguardians of our health had in mind the following consequence of what James I of England and VI of Scotland called 'A custom loathsome to the eye, hateful to the nose, harmful to the brain, dangerous to the lungs.' It happened in Portsmouth on 24th June 1809, at the height of the Napoleonic wars.

A battalion of returning soldiers had landed and stowed their baggage and ammunition on the quayside where, according to a report from Portsmouth published in *The Sussex Weekly*,

> '... they remained till this morning, when an old women emptying a pipe which she had been smoking among the baggage, the sparks fell on a barrell of gunpowder, and an instant explosion took place. The effect was most dreadful. About 30 men, women and children, were literally blown to atoms...'

Much damage was done to property. The barrel of gunpowder that had exploded stood in a row with sixteen others, which for several hours were also expected to blow up since smouldering fragments were scattered over them:-

> '... but a company of the Worcester Militia, with some resolute sailors at their head, ventured to the spot, and cleared the burning fragments from the remaining barrels. Previous to this bold enterprise, which will doubtless be duly rewarded, almost all the families fled in confusion to Portsdown Hill, expecting the whole town to be destroyed by the apprehended explosion, but they have since returned, imploring blessings upon the heads of the poor fellows who saved the town from general destruction.'

THE HAPHAZARD STORAGE OF DANGEROUS SUBSTANCES was a feature of the heady days of the Industrial Revolution, revealing a carefree ignorance of the potential for disaster. In October 1854 a six-storey warehouse in Gateshead contained 4,000 tons of sulphur as well as unspecified quantities of pyrites, coal tar, manganese, arsenic, soap, guano, nitrate of soda, rags and salt. It was an explosive mixture looking for a spark, and the spark came when a fire started just after midnight in a textile mill next door. (This fire was first spotted by some eager policemen who broke down the doors of the mill to investigate, thereby creating an inrush of fresh air to fan the flames.)

The fire spread to the warehouse. Three hours later the heated contents had given more gas than could escape through the meagre openings in the roof and the whole warehouse exploded. There were 53 deaths, among whom were sleeping residents of nearby houses who were killed when chunks of masonry burst through the ceiling. After the dense cloud of sulphur and dust had cleared it was seen that the force of the explosion had hurled burning brands across the River Tyne onto the Newcastle side, where further fires broke out and joined together into one blazing disaster area. In the absence of instant TV coverage, the railway authorities ran special excursions so that the morbid could view the ruins and even Queen Victoria and Prince Albert, returning south from Scotland, had the royal train stopped on the railway bridge so that they too could look down upon the devastation.

The provision of explosion venting, incidently, is standard practice these days but little was known of its importance when that Gateshead warehouse was built. It is of some interest, therefore, that when the Powder House of Culzean Castle in Ayrshire was built in 1880 the designer showed a gratifying degree of enlightenment and common sense. The gunpowder was needed for the firing of the Castle's gun at 8 a.m. each day on the West Green. The Powder House was built in a remote part of the estate and provided with a hollow, roofless tower, so that any accidental ignition of the powder would be vented upwards.

AN EXPLOSION WHICH DESTROYED A SHIP, the *Athol*, in 1784 was described in colourful language by one of the few survivors, the ship's doctor. He confidently pronounced upon the cause:

> '... about seven o'clock we were alarmed with the cry of fire in the lazaretto [sick bay], where the spirits are kept; the flames were already violent, and spreading rapidly: immediately under the lazaretto is the powder magazine; you can better conceive than I can describe our deplorable situation: sixty of our seamen impressed, and only the officers, with a very few who remained to suppress a dreadful fire. We exerted ourselves to the utmost,

hoisted a signal, and fired guns of distress, which soon brought great numbers to our assistance. Their efforts seemed at last to be blessed with success; the flames became moderate, and we began to think ourselves secure; fatal security to many! for in about fifteen minutes from eight o'clock the ship blew up.

'I was stunned and thrown down with the explosion, and before I could recover from the shock, a yard fell across me, attended with much excruciating torture; my sight failed me; but just as I was sinking, I recovered so far as to cling to the spar which was above me on the quarter-deck, owing to all the spars which are placed along the middle of the ship being thrown upon it. I was carried on board the *Juno* frigate, where I was treated with the greatest kindness and humanity, and am now thoroughly recovered.

'The cause of this dreadful affair was the villainy and carelessness of our cooper and steward, who were employed in stealing liquors; they had stuck a candle against a beam, which dropping into the bucket full of spirits, immediately set it on fire, as also the puncheon; they attempted to smother it by putting in the bung, but it instantly burst the calk and threw the burning spirits all over the lazaretto, which was full of spirits, oil, pitch, and cordage, being only separated by the deck from the magazine, which was directly under it.'

A total of 144 persons were killed in that explosion.

WHAT WAS PROBABLY THE FIRST OIL TANKER EXPLOSION occurred in 1886. The *Petriana* was built as a general cargo ship in 1879, but in 1886 she was modified with six tanks to carry oil. In December of that year she arrived in Liverpool to offload kerosene from Russia. This product had a flash point of 26 °C. The cargo was discharged and the ship went to Birkenhead for repairs. While the forward ballast tanks were being tested there was an explosion in a cargo tank where ten men were working. Four men died immediately and six later from the burns they received.

The accident was investigated by Sir Boverton Redwood, who found that the 'noiseless explosion' occurred when kerosene entered the tank as a spray through a defective rivet and was ignited by the candles being used by the men to illumine the tank they were cleaning. The ship was repaired and continued carrying oil until she grounded in 1903 and was abandoned.

AN UNUSUAL EXPLOSION happened in Turin in the summer of 1785, and is of note because of the scholarly interest it attracted at the time. The summer in the Piedmont area of northwest Italy that year had been a very dry one and no rain had fallen for five or six months. Signor Giacomelli was not surprised, therefore, at the dryness of the flour he had bought for his bakery in view of the extraordinarily dry corn that had been harvested.

His baker's shop had a storage room or silo at the rear and contained some 300 sackfuls of flour which had been poured into it. In the side of this silo was an opening about 3 ft square, through which the flour was conveyed into a bolter for sifting. It was already dark by 6 o'clock in the evening of the 14th December 1785, and the two boys still working were using candles. One boy was pulling flour out of the silo into the bolter. He was having some difficulty with the flour bridging and was digging about deeply into the silo to get the flour to fall.

A considerable quantity of flour suddenly fell through, forming a thick cloud which the candle on the wall ignited. There was a loud explosion in the small room where the boy was working, and the flames passed down into the shop. He suffered burns but was able to return to work two weeks later. The other boy had been working above the silo and, thinking the shop was on fire, jumped down and broke his leg.

An account of this dust explosion was given by Count Morozzo in *The Repertory of Arts and Manufactures*, Volume II, 1795. He mentioned several other instances of fire caused by candle flames igniting flour being poured, but added that no previous explosions were known. The Count attributed this particular incident to the release of 'inflammable air' formed by fermentation; the inflammable air was expelled into the room where the flour fell and the gas was then ignited by the candle. Count Morozzo's final comments are quoted below as they may be the first report proposing a loss prevention approach. They are as relevant today as they were two hundred years ago:

> 'Ignorance of the fore-mentioned circumstances, and a culpable negligence of those precautions which ought to be taken, have often caused more misfortunes and loss than the most contriving malice; it is therefore of great importance that these facts should be universally known, that public utility may reap from them every possible advantage.'

'INFLAMMABLE AIR' was also implicated in the following pit explosion in January 1794, when

'A melancholy accident happened in a coalpit, near Newcastle, by holing an old waste, charged with inflammable air, which instantly took fire from one of the workmen's candles. An explosion ensued, by which were killed 25 men and boys, and 16 horses.'

Disasters such as this often led to significant scientific advances. An explosion on 25th May 1812 at the Felling Colliery near Jarrow caused the death of 92 miners, and in 1815 Sir Humphry Davy was invited by the recently formed Society for Preventing Accidents in Coal Mines to study the causes of the explosion and devise preventative measures. By the end of the year, in a paper presented to the Royal Society, this prolific genius not only identified methane gas as the cause of the Felling explosion, but established the principles of ignition temperature, the limits of flammability, dilution limits to render a gas non-flammable, and the wall quenching of flames. His well known safety lamp was adopted in 1816.

There is a song about 'The Men of Felling' in chapter 14.

Possibly the most serious eruption of firedamp that has ever happened took place in 1879 in L'Agrappe Colliery in Belgium. One hundred and twenty-one men were killed and many more were injured. It has been calculated that in the space of three hours 12 million cubic feet of gas escaped. It ascended the shaft and ignited at the pit head, destroying all the buildings in the vicinity. The fire was extinguished by an explosion which was repeated five times at intervals of ten minutes. The last explosion was the most serious and took place after a quiet interval of nearly one hour.

MEANWHILE, GUNPOWDER was still causing problems in the late nineteenth century. On 2nd October 1874 an explosion occurred on the Regent's Canal in London and destroyed the Macclesfield Bridge. The barge *Tilbury* was carrying five tons of black powder and six barrels of petroleum. The petroleum leaked into the cabin and was ignited by a stove, detonating the black powder. The three crew members who had so imprudently lighted a fire in the stove were killed instantly and the shock of the explosion brought panic to the neighbourhood, a panic in which animals in the nearby London Zoo played a boisterous part. The iron columns of the bridge were later salvaged and re-erected. The Bridge is sometimes referred to as the Blow-up Bridge.

The title of this chapter promised a Gunpowder Plot, and here is one where the only victim was the plotter himself. As reported on 12th April 1787:-

'On Monday morning, the 9th instant, the town of Campden, in Gloucestershire, was alarmed by a violent shock, which was at first thought to be an earthquake, but which afterwards appeared to arise from an explosion of gunpowder at a house in that town, where a person had designedly, in consequence of a disagreement between himself and one of his family, set fire to a quantity of gunpowder in the garret of his son's house, which destroyed every thing in the house, leaving it a mere shell. The misguided perpetrator was blown above one hundred yards, but no person was killed except himself.'

# CHAPTER 6

# PRECAUTION AND PREVENTION

FROM THE EARLIEST DAYS of recorded history man has had a problem with accidents, and in ancient Babylon he sought to do something about it. The 'Code of Hammurabi' was devised around 1700 BC and dealt, *inter alia*, with injuries and the duty of care. It laid down the penalties to be visited upon supervisors and managers if their workers were involved in an accident:-

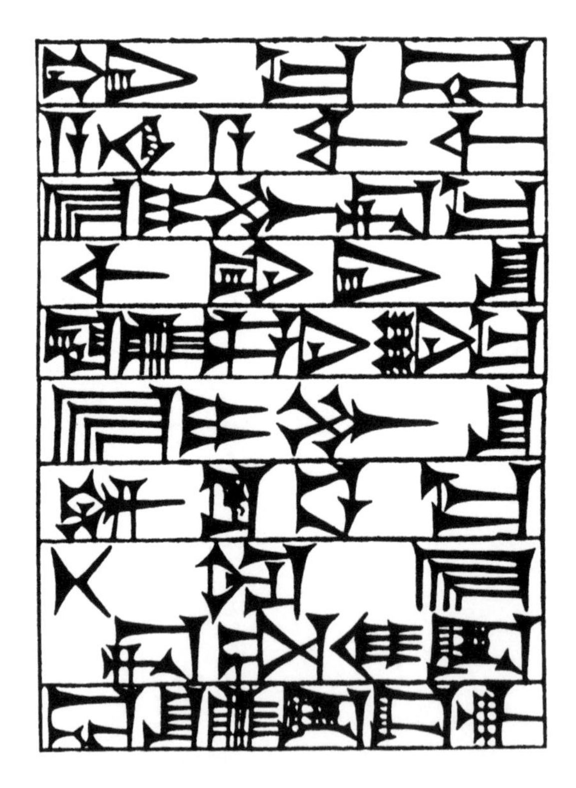

'No. 229 If a builder builds a house for a man and do not make its construction firm and the house he has built collapse and cause the death of the owner of the house — that builder will be put to death.'

The effect of this legislation on the accident rate is not recorded but we can assume that it must have jolted senior management especially into taking a fresh look at procedures in the construction industry.

The first engineering Standard comes from Old Testament times. In Deuteronomy, Chapter 22, Verse 8, we read:-

> When thou buildest a new house, then thou shalt make a battlement for thy roof, that thou bring not blood upon thy house if any man fall from thence.

FURTHER ON IN TIME, the Middle Ages also had building regulations but the aim was mainly to prevent fire spreading from one dwelling house to another. Indeed fire and pestilence were the two most common hazards of everyday life for Mediaeval man. Given the state of medical knowledge there was little he could do about disease, but early fire precautions by and large were sensible. Probably the first was the curfew; defined in the Oxford English Dictionary as a regulation in force in Mediaeval Europe, by which at a fixed hour in the evening, indicated by the ringing of a bell, fires were to be covered over or extinguished. The word comes from the old *couvre feu*, and the primary purpose appears to have been the prevention of conflagrations arising from domestic fires left burning at night. It was not very popular in England; the natives suspected the Norman Conquerors of using the curfew to prevent their gathering together when the day's work was done to grumble about their lot. It fell into disuse.

In 1189 Henry Fitz-Aylewin, first Lord Mayor of London, tried a new approach, and, in an early municipal planning bye-law, decreed that all new buildings in the city were to be built of stone, with slate or clay tiles. Owners were also obliged to have at hand large barrels of water and long ladders. Further:-

> '...he ought to make his gutter below the eaves of the said house to take the water, so that the said house shall remain safe and capable of withstanding the fierceness of a fire when it comes, and so, through it, many neighbours' houses saved and kept intact from the violence of the flames.'

In Norwich, at an 'Assembly on Friday before St. Petronilla, 28th May 1423' some fire prevention measures were laid down:-

> 'Also in every ward one person only and no more shall proclaim with kindly voice that everyone shall well and securely keep fire

and light in their places. Also all watches shall begin at the hour of curfew, to wit, at the ninth hour before midnight and shall last until the third hour after midnight... Also it is ordained that the men of every large parish within the city shall cause to be made a common ladder and two pairs of fire-hooks for pulling down burning houses. And the men of the two smaller parishes shall do the same.'

Later came the bellmen, more strident than the kindly voices of Norwich, and householders must have felt a degree of resentment at their nightly clanging. These public servants were employed under an Act of Parliament passed in 1556, and were required to patrol the streets at dusk ringing a bell and exhorting the citizens to 'Take care of your fire and candle, be charitable to the poor, and pray for the dead.'

OVER THE YEARS a formidable amount of safety legislation has been passed. Typical of the 18th century was an Act passed in the 'Second Session of the Sixteenth Parliament of Great Britain' on the 20th July 1785:-

> An Act respecting party-walls, and for the more effectual preventing mischiefs by fire; and for extending the provisions of this act, so far as relates to manufactories of pitch, etc. throughout England.

Then, of course, there is always stable-door legislation. No government is guiltless, but rarely has the failing been more vividly illustrated than in the case of the Theatres Act of 1888. Given that candle-power (and later gas) was the source of lighting it is hardly surprising that many theatres caught fire. In June 1789, for instance, it was reported:-

> 'A dreadful fire broke out, at the King's Theatre, in the Hay-market. The fire began in that part of the house, above the stage, which is called the fly. They were rehearsing a ballet below; and the first notice they had of the mischief was the sparks falling upon their heads...'

But what is probably the worst theatre fire in history occurred in September 1887, when the Theatre Royal at Exeter in Devon burned down. The fire started backstage and spread rapidly; there were few exits and there was a full house. When the fire was finally extinguished the bodies of most of the 188 dead were found huddled inside against the doors. Statistics are said to be cold indicators, but only the official table of fatalities can give a true measure of the horror of this fire.

| | |
|---|---|
| Recognisable bodies | 68 |
| Charred bodies | 46 |
| Heaps of human bones probably representing | 30 |
| Died in hospital | 9 |
| Bodies reduced to ashes among the ruins estimated in view of the list of missing as at least | 35 |
| | 188 |

The Government set up an inquiry which recommended such features as a safety curtain between the stage and auditorium, and sufficient exits clearly marked as such. The recommendations of the inquiry resulted in the passing in 1888 of the Theatres Act.

SAFETY AT SEA was also the subject of legislation, and in the great Age of Faith safety precautions could be looked on as a Christian duty. From a *Calender of Patent Rolls* for 1247 we learn:-

> 'Because it is a pious work to help Christians exposed to the dangers of the sea so that they may be brought into the haven out of the waves of the deep, it is commanded to J. son of Geoffrey, justicary of Ireland, that, so long as the lands late of W. Marshall, earl of Pembroke, in Ireland, are in his custody, he shall let the wardens and chaplains of St. Saviour's Rendenan, who are there building a tower as a signal and warning to those at sea to avoid danger, have such maintenance in money and other liveries out of the issues of the said lands as they have had, with the arrears due to them.'

Senior managers are required to formulate safety policy, and surely none more senior than the saintly King Louis IX whose ship caught fire on his return to France after the Seventh Crusade in 1254. The Seneschal of Champagne, a close companion of the King, records the incident thus:-

> 'Before we finally came to land we had another adventure at sea. One of the lay sisters in attendance on the queen was so careless that, after putting her mistress to bed, she took the kerchief the queen had been wearing round her head, and threw it down near the iron stove on which the queen's candle had been burning... the candle went on burning till its flame was low enough to set the kerchief alight, and from there the fire passed to the cloths that covered the queen's clothing.
>
> 'The queen woke to find her cabin all in flames. She jumped out of bed quite naked, picked up the kerchief and threw it all burning into the sea, and then extinguished the fire on the cloths.
>
> '...so he [the king] told them what had happened, and then said to me: 'Seneschal, I order you from now on not to go to bed till you have seen to putting out all fires, except the main fire in the ship's hold. And take note that I shall not go to bed either till you come back to tell me this has been done.' I performed this duty as long as we were at sea, and the king never went to bed till I had gone back to him.'

Similar precautions were still in force some five hundred years later. Writing at Le Havre in his diary for 5th March 1774, Robert Morris noted:-

> '... No Fire nor Candle allowed in any of ye Ships either day or night: 500 livres penalty; reason is that ye Ships are left dry and some burning ye rest cd not be removed. There is no river at

Havre... No Fires or Candles allow'd at Amsterdam when ye ships are froze up: None at Hamburg at any time after 8 in ye evening.'

THE PERCEIVED DANGERS of mining also brought legislation. In 1480 Edward IV issued a charter governing mining in the Mendip Hills (then known as the 'Myne deeps'). Here is one of the clauses which may be copied by any regulatory body that wishes:-

> Item. That if any man by the means of this doubtful and dangerous occupation do by misfortune take his death, as by falling the earth upon him, by drowning, by stifling with fire (or otherwise as in times past may have been), the workmen of the occupation are bound to fetch the body out of the earth and bring him to Christian burial at their own proper costs and charges although he be three score fathoms under the earth as heretofore.

Then there are explosions to be legislated for. An early version of the Control of Industrial Major Accident Hazards regulations was passed on the 1st December 1741:-

> 'An Act for preventing the Mischiefs which may happen by keeping Dangerous Quantities of Gunpowder in or near the Cities of London and Westminster. ...it is enacted That it shall not be lawful to have or keep more than two hundred pounds of gunpowder at any time in any house, Storehouse, Warehouse, Shop, Cellar, or other places under one or the same roof, or in any yard or yards within the Cities of London or Westminster, or within the suburbs thereof, or within three miles of the Tower of London, or within three miles His Majesty's Palace at St. James...'

THE FIRST USE OF IRON in the construction of bridges was at Coalbrookdale in 1779 and in the early days of iron bridges there were some spectactular failures. It was to be expected, then, that measures were taken to test their reliability, among them the use of elephants, since it was thought the creatures had a strongly-developed sense of danger and would refuse to cross an unsafe bridge.

Less exotic animals were used to test the Union Chain Bridge near Horncliffe in Berwickshire. It was one of the first chain suspension bridges and spans the River Tweed, the border between England and Scotland.

Designed by Captain Samuel Brown, R.N., it was opened in 1820 after being tested by twelve horses dragging carts loaded with stone.

But there is also a rule, frequently scorned as a superstition, that marching troops should break step when crossing a suspension bridge. Two incidents have been adduced as originating this rule. The English one has it that well-drilled troops marching over the bridge at Broughton, near Manchester, set up so rhythmic a motion that a pin in one of the suspension chains broke and the bridge collapsed. The French claim to be the begetters of the rule resulted from the collapse of the suspension bridge over the River Maine at Angers, when five hundred soldiers were marching across. Half of them were drowned. Certain it is that the hazards presented by a swaying bridge are well known. The designer of the Brooklyn Bridge in New York had a notice erected by the walkways to the towers of the bridge bearing the following warning:-

> Safe for only 25 men at one time. Do not walk close together, nor run, jump or trot. Break step!
>
> Chief Engineer

ROAD SAFETY needed attention very early on. In 1540 Francis I of France noted the perils of making U-turns on busy highways. He decreed:-

> 'And, under the same penalties, We forbid wagoners and drivers, whether of carts, drays, wagons, or other vehicles, to turn in the

streets, but they are to turn at the intersections and corners of said streets, to avoid the inconviences that might arise, such as wounding children or other persons and interfering with other passers-by along the road.'

The Mail Coach Service in 18th Century England was also alive to the frequency of mishaps on the roads, and issued one of the first instructions on the reporting of accidents; it could well serve as a model for present day companies. A printed circular addressed to the mail coach guards, dated 29th August 1792 and signed by Superintendent of Mails Thomas Hasker, reads:-

'When any accident happens, write what it was on the time-bill, at the very next Stage, and the next day give a more particular description of it, by Letter to me — how it happened, the cause, what was broke, and what damage was done: mind and do not neglect this.'

But in 1839 a worried Postmaster General, Lord Lichfield, is said to have opposed the introduction of the Penny Post on the grounds that the Post Office buildings would collapse under the weight of the enormous quantity of letters the cheap postage would engender.

**IN 1822 WILLIAM STUART**, Archbishop of Armagh, was accidently poisoned by his wife. The Hon. Mrs. Calvert, a contemporary, wrote in her diary on 1st May:-

'A dreadful thing happened last Monday. The Primate of Ireland had not been well, and was ordered a draught, which was to be taken directly it arrived. At the same time a bottle of laudanum was sent for one of the servants who had a bad leg. Instead of the draught, the laudanum was brought by mistake to Mrs. Stuart, who at once gave it to her husband. He died in consequence at five that afternoon, in spite of three doctors and all their remedies.'

It is said that after this fatal accident all mixtures likely to be poisonous were ordered henceforth to be put in coloured glass bottles.

**ON THE MORE GENERAL QUESTION** of the prevention of 'violent deaths', the self-improving zeal of the Victorians is to be commended. *The Insurance Guide's Handbook 1857* confidently states:-

> 'It has been predicted , and it certainly is to be hoped, that by improvement in the education and industrial training of manual labourers, rendering them more discreet; by improvement in the arts, rendering processes and engines more safe; and by moral, religious, and physical training of intellectual labourers, rendering them less liable to mental aberrations and suicides, violent deaths may be somewhat reduced.'

Governments are dab hands at banning and controlling, but it takes an individual to come up with the actual safety devices needed to help the legislation along. It is hard to believe, for instance, that the Roman Senate in the first century thought up the use of bladders that minium refiners wore over their faces to avoid inhaling dust. Pliny the Elder referred to these primitive dust masks in his encyclopaedia of natural science. 'Minium' is the Latin word for cinnabar (red mercuric sulphide).

Pliny himself was an intelligent improvisor, for when Vesuvius erupted in 79 AD he crossed the Bay of Naples towards Pompeii to investigate and thought up another protective safety device. Lacking hard hats, he ordered his shipmates to tie pillows over their heads with scarves to avoid being brained by volcanic debris. Sadly, though, Pliny did not return but died on the spot from a heart attack.

ANOTHER IDEA for improving safety was conceived by an individual and was the outcome of his personal experience of a railway accident.

In the early days of railway travel luggage was strapped to the roofs of carriages in much the same way as it had been carried in coaching times. On this occasion, luggage on the Bristol to Northwich train caught fire when a spark from the engine chimney landed on it. The engine driver, rightly looking ahead, was unaware of the fire raging behind him, and the guard was unable to attract his attention despite frantic shouting and arm waving. A group of soldiers on the train climbed onto the roof to try to put the fire out, and were themselves oblivious of impending danger. The train entered a tunnel, there was insufficient clearance, and one of the soldiers was killed. A passenger on the train, the Rev. James Barker, was much affected by his experience as a spectator. Being intelligent and public spirited, he proposed to the railway authorities the idea of a continuous cord from the guards van to the engine driver. This was said to have been accepted and became the communication cord we know today.

The Clergy in the 19th Century does indeed seem to have taken to its

collective heart the safety of railway travel. In the 1840s the Rev. Sydney Smith was much perturbed by the early custom of locking the doors of railway carriages to prevent, so it was said, witless passengers from endangering their lives by jumping from the train while it was still moving; more probably it was to ensure that nobody escaped without paying his fare. After a tragedy on a train in France, when a large number of similarly imprisioned passengers were burnt to death, the Smith of Smiths wrote:-

'We have been, up to this point, very careless of our railway regulations. The first person of rank who is killed will put everything in order, and produce a code of the most careful rules. I hope it will not be one of the bench of bishops; but should it be so destined, let the burnt bishop — the unwilling Latimer — remember that, however painful gradual concoction by fire may be, his death will produce unspeakable benefit to the public.'

HALFWAY THROUGH the 18th century Benjamin Franklin told mankind *How to Secure Houses, etc. From Lightning*:-

'It has pleased God in his Goodness to Mankind, at length to discover to them the Means of securing their Habitations and other Buildings from Mischief by Thunder and Lightning. The Method is this: Provide a small Iron Rod (it may be made of Rod-iron used by the Nailers) but of such a Length, that one End being three or four Feet in the moist Ground, the other may be six or eight Feet above the highest Part of the Building. To the upper End of the Rod fasten about a Foot of Brass Wire, the Size of a common Knitting-needle, sharpened to a fine Point; the Rod may be secured to the House by a few small Staples. If the House or Barn be long, there may be a Rod and Point at each End, and a middling Wire along the Ridge from one to the other. A House thus furnished will not be damaged by Lightning, it being attracted by the Points, and passing thro the Metal into the Ground without hurting any Thing. Vessels also, having a sharp pointed Rod fix'd on the Top of their Masts, with a Wire from the Foot of the Rod reaching down, round one of the Shrouds, to the Water, will not be hurt by Lightning.'

George III who, despite his occasional bouts of madness, was a scientifically minded monarch, disagreed with Franklin's pointy end and maintained that the conductors should have a spherical top. As a contemporary piece of verse had it :-

*While you, great George, for safety hunt,*
*And sharp conductors change for blunt,*
*The nation's out of joint.*
*Franklin a wiser course pursues,*
*And all your thunder fearless views,*
*By keeping to the point.*

THE DANGERS OF EATING contaminated food occupy the minds of many. In 1788 a grieving one-woman pressure group inserted the following warning in a local newspaper. It was picked up by a national paper 'as a circumstance of very serious importance':-

'A CAUTION. One whose afflictions may be supposed to speak most forcibly, intreats the confectioners and gingerbread bakers who may read this advertisement, never more to use what is termed by them frosting their cakes, in order to shew their sweetmeats to greater advantage, and by this means tempt unwary children to slow but certain destruction. The shining particles which they use for this purpose are nothing but minute particles of coloured glass, whose terrible and destructive consequences have deprived me of a little cherubim.

A MOTHER'

Dangerous practices of this kind were widespread by the 19th century and in 1820 Frederick Accum, described as an 'Operative Chemist, Lecturer in Practical Chemistry, Mineralogy and Chemistry', published:-

A Treatise on Adulteration of Food
and Culinary Poisoning
Exhibiting
The Fraudulent Sophistication
OF
Bread, Beer, Wine, Spirituous Liquors, Tea, Coffee,
Cream, Confectionary, Vinegar, Mustard, Pepper, Cheese, Olive
Oil, Pickles
AND other Articles employed in Domestic Economy
AND
Methods of Detecting Them.

In 1794, one James Anderson, in the Sixth Number of the *Repertory of Arts and Manufactures*, fulminated against 'The pernicious practice of keeping milk in leaden vessels' because they:-

'..communicate to the butter and milk which has been kept in them a poisonous quality, which inevitably proves destructive to the human constitution. To the prevalence of this practice, I have no doubt, we must attribute the frequency of palsies, which begin to prevail so much in this kingdom: for the well known effect of the poison of lead is, bodily debility, palsy, — death!'

Indeed, in the 1760s a Dr. Baker had shown that a frequently fatal disease in the cider counties, called endemial colic, was in fact lead poisoning, caused by the use of lead in the apple presses.

ANOTHER INDIVIDUAL with a sufficient sense of responsibility to issue a public warning against a perceived danger was the incumbent of a church at Bilton in Warwickshire. A memorial there reads:-

'In memory of Joseph, son of Joseph and Mary Meek, who was accidently drowned in the cistern of the day school adjoining this church April 30th 1845 aged 8 years. This distressing event is recorded by the Minister, as an expression of sympathy with the parents and caution to the children of the school — a report to the propietors of the open wells, pits and landslips; the want of fencing about which is the frequent cause of similar disasters in these districts; and as a memento to all of the uncertainty of life, and the consequent necessity of immediate and continued preparation for death.'

But that other clerical friend of ours whom we met in Chapter 2, the Rev. John Skinner, could see safety precautions only in their moral dimensions. Writing in his journal in March 1816 he worried:-

'This breach of the Sabbath was in France the forerunner of the Revolution... The sacred ordinance of rest from labour seems to be entirely disregarded, for the bailiffs of the coal works without scrupple employ it in repairing any accidents which might have occurred in the week, in cleansing out the boilers of the steam engine, etc. Several times I spoke to Goold on the subject... I said it was true that any work that was necessary for our sustenance and the relief of our fellow creatures, or as an immediate prevention against accidents might be performed...'

WOMEN, OF COURSE, can be just as lacking in safety awareness as men but here is a woman who displayed an admirable approach to safe practices. She was Princess Lieven, wife of the then Russian Ambassador in London. Writing to Prince Metternich in Austria in February 1820 and much concerned for consumer safety, she says:-

'...I have set to work getting you your hydrogen gas lamp. They send such lamps here from Edinburgh; but there have been accidents and they will not hold themselves responsible if the bearer takes fire instead of the wick. I have consulted one of the chief ministers here. It is a little discouraging at the start, but I shall go on. To satisfy me, it is absolutely essential that the invention should be perfected, and I don't despair of obtaining for you the apparatus you want. Speaking of chemistry, there is much talk of a fusion of parties to form a new ministry...'

In another letter to the Prince, dated 16th October 1820, she is happy to be able to report on an early sprinkler system:-

'... Not long ago, the Duke of Wellington took me to see Woolwich Arsenal, of which he is head. ... What interested me most was a new invention for putting out fires. On pressing a spring you can flood a building in an instant by means of wheels which throw out water in every direction, producing an appearance like that of fireworks. You probably will not understand a word of my explanation about fireworks which put out real fires; but picture to yourself temples, palm-trees and whatever else you see in the Prater on Sundays, made of water instead of fire, and you have Congreve's new invention.'

Sir William Congreve had indeed invented and patented a sprinkler system in 1812.

But early in the present century women themselves were under attack. An American fire journal published a sexist warning against the new danger it had detected:-

'Women in her natural sphere may not be more careless than man; but when she oversteps its boundary and defiles her femininity by smoking cigarettes, she loses many of her innate virtues and becomes even more reckless than man. Besides, she is apt to smoke in the privacy of her boudoir, where abound all the flimsy draperies which are the chief attraction of such exclusive quarters, to which ordinary men may not obtain access. These are more susceptible to ignition than are the surroundings where men usually do their smoking.'

REVERTING FOR A MOMENT to the subject of lamps, at one time in the nineteenth century it was calculated that a quarter of all deaths by fire in London were caused by lamp oil accidents. Cheap American oil was blamed, but the government failed to legislate effectively against its import, ignoring such evidence as this epitaph on a gravestone in Girard, Pennsylvania:-

In Memory of
Ellen Shannon
aged 26 years
Who was fatally burned
March 21st 1870
by the explosion of a lamp
filled with R.E.Danforth's
Non-Explosive
Burning Fluid.

Scottish lamp oil, however, made from the roasting of Lothian shale, was more highly refined and had a higher flash point; consequently it was less volatile and much safer. With awesome confidence in its user friendly product, the Linlithgow Oil Co. Ltd. advertised around 1880:-

IMPERIAL WHITE BURNING OIL

Guaranteed not to flash under 120° Fahranheit, Abel Close Cup test. The Select Committee of the House of Commons reported that if entire immunity from danger were to be sort [sic] the Flash Point would have to be raised to 120° Fahranheit. This Flash Point we are prepared to Guarantee. By using Imperial White all danger of explosion is completely Guaranteed.

DIVINE GUIDANCE in a matter of loss prevention was sought — and apparently obtained — by Sir Thomas Fowell Buxton in December 1817. Complaints had reached his nonconformist ears that men were working in his brewery on a Sunday and he wrote in his Commonplace Book:-

'To inquire into this, in the morning I went to the brew-house, and was led to the examination of a vat containing 170 ton weight of beer. I found it in what I considered a dangerous situation, and I intended to have it repaired the next morning. I did not anticipate any immediate danger, as it has stood so long. When I got to Wheeler Street Chapel, I did as I usually do in cases of difficulty — I craved the direction of my heavenly Friend, who will give rest to the burdened, and instruction to the ignorant.

'From that moment I became very uneasy, and instead of proceeding to Hampstead, as I had intended, I returned to Brick Lane. On examination I saw, or thought I saw, a still further declension of the iron pillars which supported this immense weight; so I sent for a surveyor; but before he came, I became apprehensive of immediate danger, and ordered the beer, though in a state of fermentation, to be let out. When he arrived, he gave it as his decided opinion that the vat was actually sinking, that it was not secure for five minutes, and that if we had not emptied it, it would probably have fallen. Its fall would have knocked down our steam-engine, coppers, roof, with two great iron reservoirs full of water — in fact the whole Brewery.'

BORN OF THE EXPERIENCE of the Great Fire of London, a preoccupation of the late 17th century was how best to contain a fire. One way was to blow up surrounding properties. A contemporary writer, Narcissus Luttrel, describing a fire in January 1679 at the Middle Temple in London, noted: '... but the chief way of stopping the fire was by blowing up houses, in doing which many were hurt...'.

Luttrel also wrote of a fire at Wapping on 19th November 1682:-

'It burnt all that night, and the next day till about seven in the evening, in which time it consumed near 1,000 houses. There being a great wind it burnt most furiously, notwithstanding the playing of several engines and the blowing up of divers homes... There were many persons blown up and killed. Some say forty or fifty; others more.'

Other methods of dealing with fires have been tried. In 1791 a fire broke out at an iron foundry in London and rapidly spread to an adjoining

timber yard. In the absence of anything else '... the engines were supplied with beer from a large storehouse upon the premises.' The beer was effective and there were no casualties.

That fire controllant, probably the forerunner of the foam extinguishing systems, was not widely thought to be very practical and is not in general use today. A better substance has been found, and the following report (attributed to Mr. Norman Mischler, Chairman of the UK branch of Hoechst and reproduced by kind permission of the *Hazardous Cargo Bulletin*) is of particular interest since it well illustrates the propensity — one could almost say death wish — of pressure groups in holding up the adoption of new safety precautions.

> 'Imperial Chemical Industries Limited has announced the discovery of a new fire-fighting agent to add to their existing range. Known as WATER (Wonderful and Total Extinguishing Resource), it augments, rather than replaces, existing agents such as dry powder and BCF which have been in use from time immemorial. It is particularly suitable for dealing with fires in buildings, timber yards, and warehouses. Although required in large quantities, it is fairly cheap to produce.
>
> 'It is intended that large quantities of about one million gallons should be stored in urban areas and near other installations of high risk. BCF and powder are usually stored under pressure, but WATER will be stored in open ponds or reservoirs and conveyed to the scene of the fire by hoses and portable pumps.
>
> 'ICI's new proposals are already encountering strong opposition from safety and environmental groups. Professor Connie Barriner has pointed out that if anyone immersed his head in a bucket of WATER, it would prove fatal in as soon as 3 minutes. Each of ICI's proposed reservoirs will contain enough WATER to fill half-a-million two gallon buckets. Each bucketful could be used 100 times so there is enough WATER to kill the entire population of the United Kingdom. Risks of this size, said Professor Barriner, should not be allowed whatever the gain. What use was a fire-fighting agent that could kill men as well as fires?
>
> 'A local authority spokesman said that he would strongly oppose planning permission for construction of a WATER reservoir in his area unless the most stringent precautions were followed. Open ponds were certainly not acceptable. What would prevent people falling in them? What would prevent the contents from

leaking out? At the very least the WATER would need to be contained in a steel pressure vessel surrounded by a leak-proof concrete wall.

'A spokesman from the Fire Brigades said he did not see the need for a new agent. Dry powder and BCF could cope with most fires. The new agent would bring with it risks, particularly to firemen, greater than any possible gain. Did we know what would happen to this new medium when it was exposed to intense heat? It had been reported that WATER was a constituent of beer. Did this mean that firemen would be intoxicated by the fumes?

'The Friends of the World said that they had a sample of WATER and found it caused clothes to shrink. If it did this to cotton, what would it do to men?

'In the House of Commons yesterday, the Home Secretary was asked if he would prohibit the manufacture and storage of this lethal new material. The Home Secretary replied that, as it was clearly a major hazard, local authorities would have to take advice from the Health and Safety Executive before giving planning permission. A full investigation was needed and the Major Hazards Group would be asked to report.'

# CHAPTER 7

# CHEMISTS AND ALCHEMISTS

IT IS AN INTERESTING FACT that whereas astrology remains with us long after branching out into the more respectable science of astronomy, alchemy sank without trace once scientific chemistry took over. And of course it is easy to laugh at the idea of the Philosopher's Stone which would solve all problems, but the transmutation of elements has been less of a joke since 1919, when Lord Rutherford achieved it. In 1941 researchers artificially produced gold from mercury, but this great dream of the alchemists is a non-starter; it would cost some millions of pounds to produce just one pin head of gold.

THE FOLLOWING ACCOUNT of the explosions that can occur in the practice of alchemy is taken from The Canon's Yeoman's Tale from *The Canterbury Tales* by Geoffrey Chaucer, written about 1386, and translated into modern English by Nevil Coghill. According to the translator, the details of alchemical techniques described by Chaucer are accurate and reliable in so far as they can be checked. Some writers think he had first-hand knowledge of alchemy.

*It happens, like as not,*
*There's an explosion and good-bye the pot!*
*These metals are so violent when they split*
*Our very walls can scarce stand up to it.*
*Unless well-built and made of stone and lime,*
*Bang go the metals through them every time*
*And some are driven down into the ground*
*— That way we used to lose them by the pound —*
*And some are scattered about the floor;*
*Some even jump into the roof, what's more.*

*Some said the way the fire was made was wrong;*
*Others said, 'No — the bellows. Blown too strong.'*
*That frightened me, I blew them as a rule.*
*'Stuff!' said a third. 'You're nothing but a fool,*
*It wasn't tempered as it ought to be.'*

*'No!' said a fourth. 'Shut up and listen to me;*
*I say it should have been a beech-wood fire*
*And that's the real cause, or I'm a liar.'*
*I've no idea why the thing went wrong ;*
*Recriminations though were hot and strong.*
*'Well,' said my lord, 'there's nothing more to do.*
*I'll note these dangers for another brew;*
*I'm pretty certain that the pot was cracked,*
*Be that as may, don't gape! We've got to act.*
*Don't be alarmed, help to sweep up the floor*
*Just as we always do, and try once more.'*

Clearly the practice of alchemy had its share of hazards. In 1591 one Marco Bragadino was sentenced and hanged after failing to transmute to gold some money given him by the Duke of Bavaria. And in May 1653 John Evelyn recorded in his famous diary:-

> 'My servant Hoare... fell of a fit of apoplexy, caused, as I suppose, by tampering with mercury about an experiment in gold.'

Evelyn regarded alchemists as con-men, as indeed many of them were. In 1664 he was visiting Italy, and wrote at Florence:-

> '... they showed me an iron nail, one half whereof being converted into gold by one Thurnheuser, a German chymist, is looked on as a great rarity; but it plainly appeared to have been soldered together.'

This was a simple piece of deception. A nail was fabricated, half of iron and half of gold (soldered, as Evelyn had spotted) and covered with some black substance. The gold end was then dipped into a liquid and stirred, whereupon the black was washed away and lo! the magic liquid had turned the dipped half into gold.

BUT SERIOUS DANGERS faced the serious alchemist. Scientists were evidently as distrusted by the layman in 1449 as they are, regretfully, today. Robert Bolton of London had to petition the Crown for protection and King Henry VI kindly granted him a licence to practice his chosen profession:-

> 'The King to all to whom these presents shall come, greeting.
> 'Know that since our beloved Robert Bolton has shown to us by a certain petition that although he wishes to work on certain materials by the art or science of philosophy, that is to say, to change imperfect metals from their own kind and by the same art or science to transubstantiate them into perfect gold or silver to remain and last by all proofs and standards like gold and silver, growing into these materials; nevertheless certain persons, malevolent and ill-wishing towards him, suppose him to be working by the illicit art, and so could hinder and disturb him in the practice of the art or science.
> 'We considering the foregoing and wishing to know the conclusion of the work or science, of our special grace we concede and give our licence to Robert, by whatever name he is called, that during his lifetime he may work, practice and exercise the art or science, lawfully and safely, without denunciation, hindrance, molestation or any accusation of us...'

It will be noted that the King was not a disinterested spectator of the scientific scene but had caught a glimpse of future benefits if Master Bolton's experiments were successful.

Another monarch with an eye to the main chance was Augustus II, Elector of Saxony in the 16th century. He was a great promoter of education, art, agriculture and commerce, but he was also an enthusiastic alchemist. He set one David Benter to work on the production of gold and when he failed Augustus threw him into prison. Benter defiantly inscribed on the wall of his dungeon: 'Caged cats catch no mice', which so took the Elector's fancy that he released Benter and put him to work again. But Benter still did not manage to transmute anything into gold and eventually poisoned himself.

JOHAN BAPTISTA VAN HELMONT, born in 1579, was an interesting figure, who also had a foot in both chemical camps. He believed in the aspirations of the alchemists and narrated his own transmutation of mercury into gold with the aid of a chip of philosopher's stone given him by a stranger. But he also broke new ground in chemistry and coined the word gas, almost certainly derived from the Greek *chaos*. He nearly killed himself when sixty-five by inhaling the poisonous fumes of burning charcoal.

*Gas* would have been one of the new words referred to by William Gilbert, a physician to Queen Elizabeth I, in his *De Magnete*, published in 1600:-

> 'Therefore we sometimes employ words new and unheard-of, not (as alchemists are wont to do) in order to veil things with a pedantic terminology and to make them dark and obscure, but in order that hidden things that have no name and that have never come into notice, may be plainly and fully published.'

WE WONDER WHAT Gilbert would have made of the vast number of chemical compounds manufactured today and the problem of finding 'new and unheard-of' names for them all. Confusion abounds, as one of many incidents will show. An unpleasant but useful chemical called tricresyl phosphate was commonly known in the industry as TCP. But then an antiseptic was marketed under the brand name TCP (short for trichlorophenylmethyliodisalicyl), and one inexperienced worker at a chemical plant dabbed the contents of a drum labelled TCP on his cut finger. He needed hospital treatment, so severe were the effects of what has now been renamed tritolyl phosphate.

A very early type of chemical warfare was practised by the ancient Britons, albeit for defensive purposes. Gerald of Wales, writing about 1188, quoted Julius Caesar on 'the Britons':-

> '... Julius also adds, that the Britons, previous to an engagement anointed their face with a nitrous ointment, which gave them so ghastly and shining an appearance, that the enemy could scarcely bear to look at them, particularly if the rays of the sun were reflected on them...'

This nitrous ointment sounds very much like a concoction painted on the skin as a beauty treatment several centuries later. In 1810 Mr. N. Meredith warned against its use in his book *Rudiments of Chemical philosophy; in which the First Principles of that Useful and Entertaining Science are Familiarly Explained, and Illustrated.* Under a section called 'The Magistery of Bismuth' we read:-

> 'When the salt called nitrate of bismuth is formed by dissolving the metal in nitrous acid; it does not dissolve in water, as the other salts do, but is precipitated in the form of white powder, this is the white oxyd formerly called magistry of bismuth; its beautiful white has occasioned its use, by the ladies, as a paint for the skin; but, not to mention the bad effects of almost all substances rubbed on the skin, in stopping up the pores, a lady should be very cautious in using this beautifier, lest, by exposure to the putrid effluvia, or sulphuretted hydrogen, such as the Harrowgate water abounds with, or even by sitting too near the fire, her lovely white complexion should be suddenly turned to a copper brown; an instance of which is related by Mr. Parkes in his Chemical Catechism.'

That particular skin preparation was probably a less hazardous choice than some of the other substances that have been used by beauty-conscious ladies. It is said the early Egyptians found crocodile droppings effective when used as a face-pack, while Queen Elizabeth I and her contemporaries used white lead to give their complexions the fashionable pale look. The women of Carpathia in Czechoslovakia were at one time noted for the beauty of their pearly skins, a result achieved by taking small but frequent doses of arsenic from their youth.

Such skin care could have fatal results. In 1779 Fanny Burney noted in her famous diary:-

> 'A new light is thrown upon the death of poor Sophie P. Dr. Hervey of Tooting, who attended her the day before she expired, is of opinion that she killed herself by quackery, that is, by cosmetics, and preparations of lead or mercury, taken for her complexion, which indeed was most unnaturally white. He thinks therefore, that this pernicious stuff got into her veins and poisoned her. Peggy P., nearly as white as her sister, is suspected strongly of using the same beautifying methods of destroying herself...'

FEMINISTS ARE SAID to despise any aids to beauty, however safe, but they will be cheered by the knowledge that the first woman chemical engineer was in at the very start of the profession. She was Mary the Jewess, an Egyptian alchemist of the early Christian era. She is said to have pioneered the distillation of crude oil which had seeped to the surface in what are now Iran and Iraq, and to have invented the water bath for heating glass flasks. A similar appliance is used in cookery, called a Bain-Marie from the Latin *balneum Mariae* (bath of Mary), but whether this is derived from the lady alchemist's apparatus is disputed.

In the sixteenth century appeared the Iatrochemists, devoted to the application of chemistry for medical purposes — pharmacists in fact. One of them is known to history as Paracelsus (1493–1541) a Swiss chemist of violent temper, physician and professor of physics and surgery. His real name was Theophrastus Phillipus Aureolus Bombastus von Hohenheim. He pointed out the fallacies of many medical treatments and preferred the use of specific chemical remedies instead of indiscriminate bleeding and purging. He is said to have burned the works of Galen and Avicenna in a pan of sulphur and nitre expressing the hope that their departed authors were suffering a like fate. He was lucky the mixture did not explode and send him to join his hated authors. He died in 1541, some allege in a state of intoxification.

Basile Valentine, a Benedictine monk, is the supposed author of a quaintly titled book *The Triumphal Chariot of Antimony*, first published in 1604. In it he describes how he found that the appetites of the pigs in his care had improved after being fed some compounds of what was then known as stibium. Knowing the similarity between these compounds and arsenic, long in use as a medical aid, he sought to fatten up his fellow monks after a period of fasting by dosing them with the same stuff the pigs had thrived on. The monks did not thrive, and many of them died. It has been said that this unhappy outcome gave 'stibium' its new name of antimony, from the French *anti-moine* or 'monk's bane'. The word *stibium* had been derived from the Latin for a mark and refers to its use in ancient times as eye make-up. (The Douay edition of the Bible relates that Jezebel 'painted her face with stibic stone and adorned her head.') Certain it is that our element antimony is symbolised in chemistry by the letters Sb.

The latest theory about the early death of Mozart, incidentally, is that he was not poisoned by Salieri, as is often said, but that he died from medical overprescription of antimony. If so, Iatrochemistry has a lot to answer for.

SHORTLY BEFORE THAT 1791 tragedy, an accident was reported in the *New Annual Register* on 8th July 1785 which pointed out another hazard of dealing with medical chemists, at this period known as apothecaries:-

> 'St. Austel, Cornwall. Last Sunday Mr. Avard, five of his children and two boarders, having with their dinner drunk some table-beer which had been poured out of a bottle, were in a few minutes taken very sick and began to vomit. Mr. Grant, surgeon, was immediately sent for, and after having observed the symptoms, declared that they were poisoned. He examined the bottle out of which the beer was poured, and found near two tea-spoonfuls of calx of mercury, which stuck to the bottom. Mr. Grant's endeavours for the recovery of the unhappy sufferers were, notwithstanding, attended with such success, that after their suffering great pains at intervals for three or four days, the poison was happily expelled, and they are all apparently in a fair way of recovery.
>
> 'The bottle was supposed to be bought several years ago at the sale of the goods of Mr. Warrick, surgeon, deceased; and though it had been washed out before the small beer was put therein, yet the poison stuck to the bottle, unobserved by the person who washed it. It is hoped that this accident will be a caution to people how they use old bottles, particularly such as may have been bought at druggists or apothecaries.'

Robert Boyle, born in 1627, is generally regarded as the founder of modern chemistry and in 1661 he published a book with the catchy title *The Sceptical Chymist: or Chymico-Physical Doubts & Paradoxes, touching the*

*Spagyrist's Principles commonly call'd Hypostatical, as they are wont to be Propos'd and Defended by the Generality of Alchemists.* He investigated gunpowder, and which good chemist has not, and wrote of an experiment charmingly described as 'an endeavour to fire gunpowder in vacuo with the sun beams.'

GUNPOWDER played a part in the fruitful life of the French chemist Antoine Laurent Lavoisier. For a period he was a government inspector of powder and saltpetre, but the greatest hazard of his career was the French Revolution, which chose to send him to the guillotine in 1794. He would have known of the experiments being undertaken by a compatriot who had worked in the laboratory of Pierre Joseph Maquer, for in August 1783 there was a report in the English press that:-

> 'A chemist at Paris, who had worked in the laboratory of Monsieur Macquer, employed himself in mixing experiments on gun-powder, to see if he could not add very much to the strength; he nearly ruined himself in this pursuit, but at last, observing that the state of the air had a great effect upon his operations, he tried the materials in the exhausted receiver of an air pump, into which he let in various sorts of factitious air, and found an incredible strength derived from agitating them in volatile alkaline air, insomuch that he made an ounce of gun-powder that had the strength of six ounces of the common. He was employed by the direction of the Secretary of State for the War Department.'

It seems a pity we are not told how he 'nearly ruined himself' — financially ? physically?

FRANCE, OF COURSE, was not the only nation preoccupied with the manufacture of gunpowder for warlike purposes. England, too, had a problem. Saltpetre was the vital ingredient, and in the seventeenth century its only source was decaying matter, such as animal and human waste, from which nitre (potassium nitrate) was extracted. Dovecotes were a happy hunting ground, and estates with dovecotes are said to have been able to pay part of their taxes in bird droppings. One Member of Parliament, in a debate of 1601, fulminated against the saltpetre men: 'They digge in dove cotes when the doves are nesting... cast up malting floors when the malt be green, in bedchambers, in sick rooms, not even sparing women in childbed, yea even in God's house, the church...'.

King Charles I even ordained in 1626 that his 'loving subjects' should:-

> '...carefully and constantly keep and preserve in some convenient vessels or receptacles fit for the purpose, all the urine of man during the whole year, and all the stale of beasts which they can save and gather together whilst their beasts are in their stables and stalls, and that they be careful to use the best means of gathering together and preserving the urine and stale, without mixture or other thing put therein... that if any person do be remiss thereof we shall esteem all such persons contemptuous and ill affected both to our person and estate, and are resolved to proceed to the punishment of that offender with what severity we may.'

Happily the liberty of the subject was in the safe hands of the Mother of Parliaments, and in 1656 an Act was passed prohibiting saltpetre makers from digging in houses or lands without the consent of the owners.

As a postscript to this unappealing aspect of a chemist's work, we note that in 1913 a factory blew up in New Jersey, USA, when a product called 'Praeposit' exploded. Its main ingredient was dried, pulverised horse manure.

BEFORE LEAVING THE CHEMISTS' involvement with military weapons, the humble conker is worth a glance. The French had discovered a use for horse chestnuts back in 1794, when it was reported:-

> 'The society of arts at Paris have discovered a method of producing fixed alkali, or potash, from the horse chestnut tree. A decree was in consequence passed on the 12th ult. ordering all the citizens to store the fruit of the horse-chestnut tree throughout the whole republic.'

And during the First World War children were encouraged to collect the nuts as part of the civilian war effort. One such collection is recorded at the Whitegate (Cheshire) Church of England Aided Primary School, when 5 cwt of them were sent to the Ministry of Propellants. It goes without saying that there were many cases of children falling from trees in their enthusiasm. Few of the young conker-gatherers would have known why the horse-chestnuts were needed, although some suspected the soldiers of shooting them at each other across the trenches. In fact, an important solvent, acetone, used in the manufacture of cordite was obtained by the fermentation of starch contained in the conkers. Maize had been the source of the starch until German submarine warfare restricted the supply.

BUT IT MUST NOT be thought that chemists were bent only on finding more efficient ways of killing their fellow-men. Some were directing their attention to the problems associated with gases, pure air, asphyxiation and the like. The learned Dr. Joseph Priestley began experimenting with gases in 1767 as a result, it is said, of becoming interested in the fermentation vats of the neighbouring brewery. Although he suffered a serious burn to his hand while heating concentrated sulphuric acid, the hazards he encountered were of a kind not normally associated with the practice of chemistry. He had many enemies, high and low, who disapproved of his political and religious opinions, and on the 14th July 1791 he was the victim of a riotous mob who had celebrated too well the anniversary of the fall of the Bastille. As reported at the time:-

> 'The anniversary of the French revolution on the 14th inst. which was celebrated without any disturbances at Dublin, Edinburgh, Glasgow, Manchester, Liverpool, Norwich, etc. was productive of fatal consequences at Birmingham... The mob next attacked the New Meeting-house (Dr. Priestley's), and, after trying in vain to tear up the seats, etc. they set it on fire, and nothing remains that could be consumed...
>
> 'Dr. Priestley's house at Fair-hill (a mile and a half from Birmingham) next met a similar fate, with the whole of his valuable library, and more valuable collection of apparatus for philosophical experiments. Here one of the rioters was killed by the falling of a cornice-stone.'

But he had discovered oxygen and wrote: 'Hitherto only two mice and myself have had the privilege of breathing it.'

It was not until 1837 that Gustaf Magnus showed that oxygen was carried by the blood stream to the tissues and returned as carbon dioxide to the lungs, to be expelled. The following year John Snow questioned the part carbon dioxide played in cases of asphyxiation and suggested the real problem was the lowering of the oxygen concentration. Even the popular press was showing concern. In April 1846 the *Pictorial Times* complained:-

> 'No buildings are more deficient in ventilation than places of worship. Air loaded with the products from the consumption of gas, oil and candles, chilling draughts from an immense surface of glass, inequality of heat, emanations from graveyards and sometimes from dead bodies under pews in the very centre of the building, and in some places the poisonous emanations of an open charcoal brazier passing from the corridor into the church, may all be observed producing deleterious effects.'

Dead bodies under pews? Surely not.

TWO FRENCH CHEMISTS, Messrs Sivel and Croce-Spinelli, were hailed as martyrs to science and buried in great style in Paris. In April 1875 they ascended in a balloon to test a breathing apparatus at high altitude. The apparatus was a simple bladder containing a mixture of air and oxygen, connected by a tube to a hand-held mouthpiece. It proved beneficial when a height of some 22,800 ft had been reached, and the aeronaut in charge of the balloon was urged to throw out ballast to ascend higher. One of the scientists, very oddly, even threw overboard a piece of his own equipment weighing some eighty pounds. The balloon shot upwards and the astronaut lost consciousness. When he came round the balloon was descending rapidly, but Sivel and Croce-Spinelli were lying dead in the bottom of the car, their faces black and blood running from their mouths. Evidently a hand-held breathing apparatus was unreliable, nor can there have been sufficient oxygen in the breathing mixture.

We have mentioned Lavoisier's unfortunate encounter with the French Revolution and this upheaval in European affairs had a side effect on chemistry, if we are to believe a somewhat tart review in England of scientific publications for the year 1792:-

> 'Of separate productions on air and the chemical nature of aerial fluids, our account must be very short. The attention of French

chemists is unfortunately diverted to less pleasing and less useful occupations, and our countrymen seem to have felt the shock, and it has contributed to render them torpid.'

One of these less useful occupations was the construction in 1794 of observation balloons to help the French army, and a young chemist, N. J. Conte, introduced a special varnish which he claimed was impermeable to the hydrogen used for inflating the balloons. In one of his experiments a glass flask exploded and he lost the use of one eye, officially deemed a war wound.

BUT IN THE SAME YEAR of that sniffy review of chemical advances, a report appeared in England of an 'important philosophical' discovery made by a Frenchman:-

'... the power of light to render the vitriolic acid altogether harmless to the human body; insomuch that a man may wash his hands in a substance that would otherwise reduce them to a cinder, with this sole precaution of setting the basin in the rays of the sun.'

In the last chapter we quoted Mrs. Calvert's diary entry on the death of an Archbishop, but some years earlier she had heard of another episcopal tragedy. She wrote on the 8th February 1814 about a son of the Bishop of Limerick and his encounter with some vitriolic acid which evidently had not been rendered harmless:-

'A shocking accident happened a short time ago to one of Edmond Knox's children, a fine boy of seven years old. The nurse had some vitriol in a glass by her bedside to apply for toothache. The child got up in the night and drank it; he died in eight and forty hours.'

In France, Henri Le Chatelier (1850–1936) had a narrow escape when trying to make ammonia by compressing hydrogen and nitrogen into an autoclave and heating in the presence of iron. Some air had been drawn into the autoclave through an open petcock and there was a violent explosion which shattered the vessel and flung pieces of apparatus around the laboratory. Disheartened, he did not proceed with the process, and who shall blame him.

FROM HUNGARY COMES the report of some novel recruits to the chemical industry. After the First World War a nitrocellulose factory took on a graduate chemist from Vienna to improve the yield of the process. The earnest young man carefully studied all aspects of the production process and concluded that much of the product was being lost during the nitration process when it was washed to remove surplus acid. These washings were run off into an adjacent pond and the chemist decided it would be sufficient to wash the product until the pH of the washings was down to 5. He explained it all to the foreman who, unimpressed, said he knew nothing of these new fangled ideas about pH but he did know that the product was sufficiently washed when the ducks returned to the pond.

It is unlikely that such pollution of a duck pond would be allowed today, but the chemist's confrontation with the environment is not new. Green campaigners and NIMBY activists have indeed been around for a long time; it is said, for instance, that in the 18th Century one of the first sulphuric acid plants was forced to relocate south of the River Thames from Twickenham because residents there objected to the pollution it caused.

IN ABOUT 1700 Bernardino Ramazzini reported a local environmental issue in Italy in his treatise *De Morbis Artificum* (Diseases of Workers). A translation is worth quoting as it shows how very little attitudes and procedures have changed:-

'A few years ago a violent dispute arose between a citizen of Finale, a town in the dominion of Modena, and a certain business man, a Modenese, who owned a huge laboratory at Finale where he manufactured sublimate. The citizen of Finale brought a lawsuit against this manufacturer and demanded that he should move his workshop outside the town or to some other place, on the ground that he poisoned the whole neighbourhood whenever his workmen roasted vitriol in the furnace to make sublimate. To prove the truth of his accusation he produced the sworn testimony of the doctor of Finale and also the parish register of deaths, from which it appeared that many more persons died annually in that quarter and in the immediate neighbourhood of the laboratory than in other localities. Moreover, the doctor gave evidence that the residents of that neighbourhood usually died of wasting disease and diseases of the chest; this he ascribed to the fumes given off by the vitriol, which so tainted the air near by that it was rendered unhealthy and dangerous for the lungs.

'Dr. Bernardino Corradi, the commissioner of ordnance in the Duchy of Este, defended the manufacturer, while Dr. Casina Stabe, then the town-physician, spoke for the plaintiff. Various cleverly worded documents were published by both sides, and this dispute which was literally 'about the shadow of smoke', as the saying is, was hotly argued. In the end the jury sustained the manufacturer, and the vitriol was found not guilty. Whether in this case the legal expert gave a correct verdict, I leave to the decision of those who are experts in natural science.'

THE HUNGARIAN DUCKS we mentioned above must have had an in-built feel for chemistry. Not so human beings, who have to knuckle down to years of hard study. It is advisable to begin at an early age and young chemists in the thirties had great scope for blowing themselves up at hardly any cost. Bicycle lamps in those pre-battery days needed calcium carbide, which could be bought easily from cycle shops. The trick was to put some calcium carbide into a screw-top bottle with a little water, replace the screw top and hastily chuck it into the pond. Acetylene would be produced and in a short time would decompose, producing a very satisfying explosion.

Concern for the budding chemist in the classroom prompted someone to adapt Thomas Grey's *Ode on a Distant Prospect of Eton College*:-

*Alas regardless of their doom,*
*The little victims play!*
*No sense have they, so bangs will come*
*With jets of acid spray.*

# CHAPTER 8

# GRAVE TALES

THOMAS GREY got it about right when he wrote his Elegy in a country churchyard:-

*Let not Ambition mock their useful toil,*
*Their homely joys, and destiny obscure;*
*Nor Grandeur hear with a disdainful smile*
*The short and simple annals of the Poor.*

Epitaphs and other monumental inscriptions are of great value to social historians looking into the lives and deaths of those who never made it into the history books. For our purposes too they are of interest, since they record many of the accidents from which ordinary people died. Some of the inscriptions have been rendered partly illegible by time; some may strike us as crude in their explicit detail or extravagant in their sentimentality:-

*Yet e'en these bones from insult to protect*
*Some frail memorial still erected high,*
*With uncouth rhymes and shapeless sculpture deck'd,*
*Implores the passing tribute of a sigh.*

Loughborough seems to have been a dangerous place to live in. The following have all been noted in the Parish Church of All Saints:-

Sacred to the Memory of Richd Palfreman and Elizh.
Relict of Richd.....Also John Palfreyman, eldest son ....
By accident was called out of time into eternity,
January 4th 1837, Aged 31 years.
You readers all both old and young
Consider how soon we was gone.

Near this stone are interr'd the Remains of Edward Goodman, late of this town, Carrier, who by accident was suddenly depriv'd of existence on the 1st October 1791.
***Spectator pause a moment o'er his Fate***
***In this behold thy own uncertain state***

*Of Death's approach thou mayst no warning have*
*But may like him be hurri'd to the Grave.*

In memory of Thomas Holt who was killed by the Main Wheel at the over Mill on the 13th day of December 1769 aged 29.

*A time of Death there is you know full well*
*But when, or where, or how, no one can tell*
*For sudden Death therefore be thou prepared*
*Then thou shall surely have a good Reward.*

In Memory of Samuel son of Joseph Simmonds who died of a strain in his Ancle the 11th of November 1770. Aged 12 years.

*The surgeons tried on me their skill*
*And often times my blood did spill*
*But all was in vain for none could cure*
*Therefore the pain I must endure*
*Yet I with patience bore the same*
*For Christ my Saviour's Holy Name*
*My fellow schoolboys cast an eye*
*And view these lines as you pass by*
*I lie and sleep within this ground*
*Until I hear the trumpet sound*
*Then shall I rise and leave this dust*
*And live for ever with the just.*

In Memory of Daniel Hayes junr. He was born Decr. the 14th 1790. Received a severe accident on the 11th June and changed his mortal for an immortal state on the 11th of July 1810.

The last gives no detail of the severe accident, but more graphic is this from Durness Churchyard in Sutherland:-

*Here doth lye the bodie*
*Of John Flye, who did die*
*By a stroke from a sky-rocket,*
*Which hit him in the eye-socket.*

ANOTHER VICTIM OF FIREWORKS was Richard Coulsborne who was killed on the 1st August 1715 during the celebrations of the first anniversary of the accession of George I. His epitaph at St. Magdalene's Church, Woodstock, Oxfordshire, reads:-

*It was decreed that I should die*
*By a sky rocket in my eye*
*The first of August 'tis well known*
*That brought me to my dismall doom*
*Rejoicing the King George was come*
*Which sent me forth to my long home.*

The date tells all on this inscription from a gravestone at Minister in Sheppey, Kent, witnessing to the hazards of fireworks:-

'Here lies the body of Simon Gilker Junior who was killed by means of a rockett. November 5, 1696. Aged 48 years.'

A very well preserved gravestone exists in the grounds of Winchester Cathedral and commemorates a soldier who died in a most unusual manner:-

'In memory of Thomas Thetcher a Grenadier in the North Reg' of Hants Militia, who died of a violent Fever contracted by drinking Small Beer when hot the 12th of May 1764.
Aged 26 Years.
In grateful remembrance of whose universal good will towards his Comrades, this stone is placed here at their
expence, as a small testimony of their regard and concern.

*Here sleeps in peace a Hampshire Grenadier,*
*Who caught his death by drinking cold small Beer,*
*Soldiers be wise from his untimely fall*
*And when ye're hot drink Strong or none at all.*

This memorial being decay'd was restored by the Officers of the Garrison A.D. 1781.

*An honest Soldier never is forgot*
*Whether he die by Musket or by Pot.*

The stone was replaced by the North Hants Militia when disembodied at Winchester, on 26th April 1802, in consequence of the original Stone being destroyed.
And again replaced by The Royal Hampshire Regiment 1966.'

TRAVEL AND TRANSPORT presented dangers in the past as we have seen. Only the type of hazard has changed. In Prendergrast churchyard, Dyffed, there is the following inscription:-

*Here I lie and no wonder I'm dead*
*For the wheel of the wagon went over my head.*

And in Kendal Parish Church, Westmorland:-

*Here lies John Ross*
*Kicked by a hoss.*

A MORE DRAMATIC travel hazard was presented by the highwayman. A gravestone near the gate in Chew Magna, records a fatal shooting:-

'Sacred to the memory of Mr. William Fowler who was shot by a highway man on Dundry Hill on June 14th 1814.'

A certain Mr. Dean of Imber, Wiltshire was more careful. On the roadside between Devizes and Tilstead is a stone marking the spot where he was attacked by highwaymen. He kept a cool head and used his whip with considerable effect. He called to imaginary friends to come to his assistance in order to deceive the attackers. The inscription on the stone is as follows:-

'At this spot Mr. Dean of Imber was attacked and robbed by four highwaymen in the evening of October 21st 1839. After a spirited pursuit of three hours one of the felons Benjamin Colclough fell dead on Chitterne Down.
Thomas Saunders
George Waters and
Richard Harris

were eventually captured and were convicted at the ensuing Quarter Sessions at Devizes and transported for fifteen years. This monument is erected by Public Subscription as a warning to others who presumptuously think to escape the punishment God has threatened against Thieves and Robbers.'

A similar stone was erected were Benjamin Colclough fell dead.

THEN THERE WAS the likelihood of coming across a bull maddened by being tethered to a block of wood or 'logger' to prevent straying, as recorded in an epitaph in Abenhall churchyard, Gloucestershire:-

*As I was riding on the road*
*Not knowing what was coming,*
*A bull that was loggered and pressed*
*After me came running.*
*He with his logger did me strike*
*He being sore offended,*
*I from my horse was forced to fall*
*And thus my days were ended.*

Another mad animal that could cause death was, of course, the dog and it is fortunate that we have such strict quarantine laws. Dunking in sea water was at one time believed to prevent rabies developing after a suspicious bite. The Old Meeting House at Horningsea, Wiltshire, has a worn stone recording a death by rabies:-

'In Memory of Willm. Barnes who died of the HYDROPHOBIA: He was bitten by a Dog on the 18 of May and was Dipped in the Salt Water on the 19 and died on the 31 of July following after a few hours of strong PAROXYSMS: aged 28 years. 1820.'

A more unusual death-by-animal is recorded on the tombstone of Hannah Twynnoy at Malmesbury Abbey in Wiltshire. She was a servant employed at the White Lion Inn and was mauled by a tiger at a local circus.

'In memory of Hannah Twynnoy who died October 23rd 1703, Aged 33 years.

*In bloom of life*
*She's snatch'd from hence*
*She had not room*
*To make defence:*

*For Tyger fierce*
*Took life away*
*And here she lies*
*In a bed of clay,*
*Until the resurrection day.'*

Legend has it that she was trying to balance a bun on the nose of the tiger.

THEN THERE IS always the weather, and too much of it can be fatal. Lightning in particular has caused many deaths and is a continuing hazard. This gravestone, originally in St. Michael's Church, Stanton Harcourt in Oxfordshire, was erected by Lord Harcourt in memory of two young farm workers:-

Near this place lie the bodies of
John Hewet and Sarah Drew
An industrious young man, and
Virtuous maiden of this parish;
Contracted in marriage
Who being with many others at harvest
Work, were both in one instance killed
By lightning on the last day of July
1718.

*Think not by rigorous judgment seiz'd,*
*A pair so faithful could expire;*
*Victims so pure Heav'n saw well pleas'd*
*And snatched them in Caelestial fire.*

*Live well & fear no sudden fate;*
*When God calls Virtue to the grave,*
*Alike tis Justice, soon or late,*
*Mercy alike to kill or save.*

*Virtue unmov'd can hear the Call,*
*And face the Flash that melts the Ball.*

The verse is attributed to Alexander Pope, who certainly wrote this couplet on the tragedy:-

*Here lye two poor Lovers, who had the mishap*
*Tho very chaste people, to die of a Clap.*

Another tragic victim of lightning is remembered at Bury St. Edmunds:-

Here lies intered the body of Mary Singleton, a young maiden of this parish, aged 9 years; born of Roman Catholic parents and virtuously brought up, who being in the act of prayer, repeating her vespers, was instantaneously killed by a flash of lightning, August the 16th 1785.

And yet another, at Winchcombe in Gloucestershire:-

*Here lies Joseph, Antony Myonet's son;*
*Abigail his sister to him is come.*
*Elemental fire this virgin killed,*
*As she sat on a cock in Stanway's field.*

ICE TOO CAN BE dangerous stuff. A young boy, John Rose, is remembered at Betchworth, in Surrey:-

*Dear friends and companions all*
*Pray warning take from me;*
*Don't venture on the ice too far,*
*As 'Twas the death of me.*

In 1951, in the tower of the Parish Church of St. Michael and All Angels at Brampton, Dorset, a tablet was placed above a stone originally laid to record the death in 1776 of the young son of the Parish Clerk. His name is not known, but parish records relate that he was killed outside the church tower by a falling lump of ice. The new tablet contains the original verse, now obliterated from the old one:-

*Bless my eyes*
*Here he lies,*
*In a sad pickle*
*Killed by an icicle*

THE FAILURE OF BRIDGES is a relatively rare occurrence. It was announced that on 2nd May 1845 a clown called Nelson would be drawn in a bath tub by four geese up the River Bure at Great Yarmouth. One vantage point for this spectacle was the suspension bridge over the river. Such a crowd were standing on the south side that the chains snapped and some 400 spectators fell into the river. One hundred and forty people were drowned. The following inscription on a tombstone in the churchyard of St. Nicholas, Great Yarmouth commemorates a young victim of this calamity:-

Sacred to the memory of George H.J.Beloe. The beloved son of Louisa Beloe, who was Unfortunately drowned by the fall of the Suspension Bridge. The 2nd May 1845. Aged 9 years.

*Farewell dear son, no more I press*
*Thy form of light and loveliness,*
*And those who gazed on thy sweet face*
*Knew it to be an angels dwelling place,*
*And if that realm where thou art now*
*Be filled with beings such as thou,*
*From sin set free and sorrow freed*
*Then heaven must be a heaven indeed.*

THE WEARING OF HARD HATS is all too often ignored today but, as the following epitaphs record, being in the path of a falling object is not recommended. In the Church of St. Mary and St. Nicholas at Wilton, Wiltshire, is this inscription:-

To the memory of William Spackman
Who died by the fall of a tree
On the 23rd January 1884
Sadly regretted by all who knew him
Aged 28 years.

And in Dean Churchyard, Bolton:-

*A ponderous load on me did fall,*
*And killed me dead against this wall.*

SCAFFOLDING ACCIDENTS also occur frequently and all scaffolders should be reminded of the following tragedy, recorded on a wooden memorial in the churchyard of St. Mary the Virgin, Baldock, Hertfordshire:-

In memory of Henry George, son of Henry & Harriet Brown who Departed this Life March 20th 1861 Aged 10 Years and 10 Months.
*How soon I was cut down. When innocent at play,*
*The wind it blew a scaffold down and took my LIFE away.*

AN 18TH CENTURY AMPUTEE is remembered in the churchyard at Lindfield in Sussex:-

In memory of Richard Turner who died
November 13th 1768. Aged 21 years.

*Long was my Pain, great was my Grief,*
*Surgeons I'd many, but no Relief.*
*I trust through Christ to rise with the just,*
*My leg and Thigh was buried first.*

THE ESHANESS KIRK-YARD in the Shetland Isles reflects a sad state of affairs calling for consumer protection:-

Donald Robertson. Born 4 January 1783, Died 4 June 1842, Aged 63 years. He was a peaceable quiet man and to all appearance a sincere Christian; his death was very much regretted which was caused by the stupidity of Laurence Tulloch in Clothister who sold him Nitre instead of Epsom Salts by which he was killed in the space of 3 hours after taking a dose of it.

IN ST. JOHN'S CHURCHYARD, Devizes, stands an obelisk 15 ft high over the remains of 5 persons from a wedding party who on Sunday the 30th June 1751 were drowned in Drews Pond through their unskilled management of a cooler in which, for lack of a boat, they had ventured onto the water.

'In memory of the sudden and awful end of Robert Merril and Susannah his wife, Eliz. Tiley her sister, Martha Carter and Joseph Derham who were all drowned in the flower of their youth in the pond near the town called Drews on Sunday evening the 30th June 1751 and are together underneath entombed.'

THE ONLY THING we know about this fall, recorded in Normanton Churchyard, Leicestershire, is that it was sudden.

*He died! how startling was his sudden fall*
*He's gone obedient to th'Almighty's call;*
*Dropt in a moment insensible of fear*
*No thought disturb'd Him, no mistrust was near.*

NOT TO BE IGNORED are supernatural hazards, as demonstrated by the inscription on the stone cross in the centre of Devizes Market Place:-

'The Mayor and Corporation of Devizes avail themselves of the stability of this building to transmit to future times the record of an awful event which occurred in this market place in the year 1753, hoping that such records may serve as a salutory warning against the danger of impiously invoking Divine Vengence or of calling on the Holy Name of God to conceal the devices of falsehood and fraud. On Thursday, the 25th January, 1753, Ruth Pierce, in the County, agreed with three other women to buy a sack of wheat in the market, each paying her due proportion towards the same. One of these women in collecting the several quotas of money, discovered a deficiency, and demanded of Ruth Pierce the sum that was wanting to make good the amount. Ruth Pierce protested that she had paid her share, and said . . . 'She wished she might drop down dead if she had not.' She rashly repeated this awful wish; when, to the consternation and terror of the surrounding multitude, she instantly fell down and expired, having the money consealed in her hand.'

The coroner found no marks of violence on her. It has been suggested that she probably did pay her share and that the words 'having the money concealed in her hand' were added later to embellish the story.

AT THE CHURCH OF ST. MARY at East Bergholt, Suffolk, is a tablet inscribed:-

'In Memory of Mr.John Mattinson, born in Long Sleddale near Kendale, Westmoreland. He was Eleven Years the Beloved School Master of this Town, and then Unfortunately Shott. The 23 of November 1723, Aged 32.

*Profuit et Placuit, Miscebat et Utile Dulci*
*Discilulis Terror, Deliciaeque suie.'*

'A terror and delight to his pupils' appears to be a translation of the last four words. But how he came to be shot in this accident is not recorded — perhaps one of the pupils he had terrorised took revenge.

And in Braemore Burial Register the following is recorded:-

22 September 1799. George Davidge, aged 17 years. Shot by imprudently playing with the barrel of a gun which he knew not to be loaded.

MANY PARISH INCUMBENTS and their clerks must have had an interest as great as ours in the misfortunes of everyday life, judging by the sometimes morbid detail with which they recorded the deaths of those whose burials they were entering in Parish registers. Here is a selection, poignant, bizarre, puzzling and even (to us, today) amusing.

GLOUCESTERSHIRE, in particular, has produced a fine crop of such entries:-

Parish of Cam

Charles King aged 26. Buried 5 February 1808. He died in consequence of a cart wheel going over him, about three weeks ago.

Reuben Bird aged 11. Buried 2 August 1808. His death was occasioned by the kick of a horse.

Betty Woodward. Buried 10 May 1812. Whose death was occasioned by her clothes catching fire in the absence of her friends.

John Reeves aged 25. Buried 22 April 1825. Who was accidently killed by a loaded wagon passing over his body in this parish.

James Hill aged 9. Buried 24 September 1828. Whose death was occasioned by his having accidently fallen against a wheel at the clothing factory of Mr.J.Cam.

Parish of Olveston

Betty Dyer Buried 21 March 1747/8 Who was accidently drowned in one of the Reens (a ditch).

John Curtis Buried 14 April 1785. Who died by an accidental wound from his own hatchet.

William Morgan Buried 24 October 1790. Killed by a fall from an apple tree.

Parish of Thornbury

John Hancock, Fisherman, aged 52. Buried 7 May 1777. Drown'd in the River Severne by the oversetting of his boat while fishing.

William Woodruffe, aged 19. Buried 11 July 1778. He fell upon one of the prongs of a pick, which pierced the brain.

Thomas Wilson aged 50. Buried 22 June 1779. His death was occasioned by being run down on Old Down, at a horse race.

John Whitefield, pigkiller, aged 38. Buried 21 July 1782. On the preceeding Tuesday, coming from Bristol, with his cart loaded, just without Stokes Croft Turnpike, unfortunately riding upon the shaftes, he was thrown off; the wheels went over his body, and he died the next morning, Eleven o'clock.

John Bendall aged 13. Buried 15 February 1785. This child was suffocated, by pitching over the side of a boat, and his head sticking in the mud.

Anne Pope, servant. Buried 26 March 1790. She poisoned herself by taking Arsenick.

Henry Mills aged 26. Buried 3 April 1793. Died by the visitation of God.

Hannah Hill aged 4 years 6 months. Buried 15 March 1797. She was burnt to death by her clothes accidently catching fire.

Ann Cossham, infant. Buried 30 August 1807. This child was put of bed by her parents in perfect health, and on their retiring to rest sometime afterwards they found her dead.

Caleb Lee (Butcher) Buried 10 February 1812. This man, in a state of intoxication, fell into a pool at Cowhill from whence he was with difficulty extricated, and sustained such an injury that he died within two days after.

FROM DERBYSHIRE WE HAVE:-

Parish of Eyam

Edward Torre Buried 30 June 1699. Who was killed wth a Plugg in ye Groove over against ye Parsonage Field.

From Parish of Stoney Middleton

Thomas Sellars. Buried 7 July 1718. Damped with smoce in ye Grove.

Martha Mason, daughter of Robert Mason. Buried 4 November 1737. Eliz Mason, wife of Robert Mason. Buried 8 November 1737. Those were the unhappy persons that were bloon up with gunpowder. November 2.

Parish of Dronfield

Joseph Pinedar. Buried 21 February 1766. Lost upon the moors in a remarkable qeeah snow. [We wonder if this was the same as the sticky snow reported in December 1990 or the 'wrong type of snow' blamed by British Rail in February 1991].

Ann & Martha Jenkins. Buried 19 August 1768. Both drowned or kill'd by the fall of a house in Dronfield when they were in bed — in which Downfal was occasioned by a remarkable Great Flood.

Emma daughter of Chrst Wragg. Buried 22 February 1806. Who was burnt to death by throwing shavings into a bonfire at Unstone.

THE PARISH REGISTER of Dauntsey, Wiltshire, includes a curious entry:-

Buried 1706 Apl 29. John son of Richard Hayward killed with lightening at Brinkworth, April 19 and buried on the 20th — a boy of about 14 or 15 years of age imminent for cursing and swearing, his mother used to devote her children to the devil in oaths and imprecations.

THE PARISH REGISTER of Barlaston in Staffordshire records the following tragic accident:-

19th April 1737. Jno, son of Willm and Ellen Cheadle, aged 5 unfortunately killed by a roller of a bowling green, buried.

AN ENTRY IN THE Parish Register of Stoak by Chester makes painful reading:-

24th March 1793. Richard Crimes of Whitby (in the parish of Stoak) buried. He was killed by a pike staff running up his fundament as he fell from a Hay Mow. Aged 24 years.

AND FROM THE Kirkley (near Lowestoft) Parish Register:-

14th August 1824. Groves Foreman of Kessingland, aged 28, died from falling from a corn stack upon the handle of a fork placed upright which entering the rectum empaled him.

THE PARISH REGISTER OF STRATTON ST. MARGARET, Wiltshire has the following entry:-

Burial 1793 Jly 31. John Puss. This man lost his life having been induced by a lad of 19 years old to wrestle with him, the Boy threw him and he was killed in the fall.

THE PARISH REGISTER OF BURBAGE, Wiltshire, has a puzzling entry:-

Burial on February 11 1648/9. Buried a souldier yt had bin drinking hot water and fel off his horse etc.

THE FOLLOWING from the burial register of Battle Church in Sussex tells us all we need to know:-

1764. December 5th. James Gillmore and Thomas Gillmore, both buried in one grave, who were accidently killed by the blowing up of the Sifting House at Sedlescombe Gunpowder Mills; in which house there was computed to be a ton of Gunpowder; at which time and place there was two other men killed, which were buried at Sedlescombe.

IN THE ARCHIVES OF WHITBY there is a diary which records some harrowing accidents. Here are a few of them:-

1791 A body of a child is found in a hogs head of treacle at York.

1830 Mr. Huskisson died while shaking hands with the Duke of Wellington. [cf. Mr. Creevey's report on this death in Chapter 4.]

December 12th 1830. John Brown, a dockman, whilst crossing the bridge was blown off the hearse and drowned. The horses swam to the shore and were saved.

November 26th 1853. A man when brewing at Lythe Castle fell into a boiler and came to his death.

June 28th 1857. Mrs Coffin of Barnaby Lyth was killed by a cow.

1858. Mark Brooksbanks killed on the railway lines being asleep on the rails.

October 13th 1888. At Grosmont, John Coulthirst was suffocated by gas while cleaning a stove.

PROBABLY THE SHORTEST accident record is given in Old Dalby in Leicestershire:-

1835 Henry Wells was killed.

BUT THE MOST BAFFLING must surely be the entry for 8th February 1593 in the Burial Register of Kendal in Westmorland:-

Jenne daughter of George Dobson & dyed in Kendall with jouhndipon.

# CHAPTER 9

# PACKING 'EM IN

THE EPISODE KNOWN TO BRITISH HISTORY as the Black Hole of Calcutta was hardly an accident but it is worth noting that of the 156 English prisoners herded into a cell 20 ft square in 1756 only 23 were alive the following morning. However, a tragedy of recent years has reminded us all that wherever crowds gather a potential danger is waiting in the wings. 'Waiting in the wings' is an expression taken from the theatre and it is indeed in theatres and other places of public entertainment that overcrowding and inadequate egress are most commonly encountered. A body of legislation has grown over the years to regulate such matters as, for example, the provision of emergency doors opening outward, but each new disaster requires a new lesson to be learned.

BEFORE ANY KIND OF LEGISLATION was introduced, however, accidents such as the following were commonplace. It was reported on 4th February 1794 that:-

> 'Yesterday evening a dreadful accident happened at the Little Theatre in the Hay-market. Their majesties had commanded the play, and there was a great crowd assembled before the pit door. A poor women having been thrown down, the people kept pushing forward, others were thrown over her, and all were trampled upon by the crowd, who passed over their bodies into the house. The pit lies lower than the threshold of the door leading into it: Those therefore who go in must go down a step. Here it was that the mischief happened: for the people who were the unfortunate suffers, either not knowing anything of this step, or being hurried on by the pressure of the crowd behind, fell down, while those that followed immediately were, by the same irresistable impulse, hurried over them. The scene that ensued may be easier conceived than described; the screams of the dying and the maimed were truly shocking; while those who were literally trampling their fellow creatures to death had it not in their power to avoid the mischief they were doing. One could

> scarcely have believed that so many could have been killed in so small space...'

Fifteen people were killed and some twenty suffered broken limbs. Extraordinary as it seems to us today, 'this melancholy circumstance was not generally known in the theatre till late in the evening, and it was kept from the knowledge of their Majesties till the play was over.' They evidently subscribed to the view that The Show Must Go On.

An accident of this kind used to be regarded as just one of those mishaps people brought upon themselves by their behaviour. As a mid-nineteenth century writer, commenting on the Peterloo riots in 1819, put it: 'A greater loss of life has more than once been created by a panic in a theatre, when the stronger have crushed and trampled the weaker to death, while making their escape from a real or imaginary danger.'

THE PRESENCE OF A LARGE NUMBER of children, excitable and unreasoning, is an added hazard to the problem of overcrowding, as a tragedy in 1883 shows. On 16th June of that year a special performance for children of conjuring, marionettes, etc. took place in the Victoria Hall in Sunderland. Advance publicity had suggested presents and prizes for all, and well over 2,000 expectant children turned up. Half were accommodated in the main body of the hall, the remainder in the gallery.

At the end of the show the presents were not handed out but were tossed from the stage into the audience on the main floor for the youngsters to scramble for. The children in the gallery soon realised that to receive a present they needed to get below. They raced down a flight of stairs, along a passage and down another stairway. At the foot of this second staircase was a pair of swing doors, which had been locked into a position so that only one person at a time could get through a 2 ft wide opening; this was a precaution taken by the management to ensure that each person entering had bought a ticket. It was at this 'exit' that some 400 hysterical children hurled themselves, those in front being jammed up against the door by others pressing from behind. And, as is the nature of children, some of them tripped on the stairs, adding more small bodies to the squirming pile at the foot of the staircase.

A total of 186 children were crushed to death in the Victoria Hall that day. A doctor living opposite, who was one of the first on the scene, wrote afterwards:-

> 'It was alleged by the owner of the hall that the fatal door was placed there for better securing the safety of the public on

> occasions of great gatherings. A fig for such nonsense! It was placed there, in my opinion, for securing the safety of the money taken from the public.'

A remarkably similar tragedy had occurred in Malta in 1823, when a locked door and the handing out of goodies to small children were responsible for the loss of life. The accident is the more poignant because the children were being protected in the quiet of a convent from the possible excesses of a public spectacle. Private William Wheeler of the 51st Regiment, then on garrison duty in the Island, wrote about it to his family in England. He describes the annual carnival in Valetta, with its three days of fun and jollity, and continues:-

> 'Every face seemed to smile with content, until about half an hour after the carnival had closed, when the whole city, I might say Island, was thrown into the deepest sorrow and anguish of mind imaginable.
>
> 'I will endeavour to describe the cause as clear as I am able. In Valetta there are a number of families who consider it improper that the children should be spectators of so much extravagance and folley. Each day they send them to the convent of Jesus Maria and Joseph, where they are amused and entertained by the monks until dusk, when they are let out to go home to their friends. The children have to pass down a long passage in the centre of which stands an image of the Virgin, here a few priests are stationed to distribute sugar plumbs, sweetmeats etc. to them as they pass. To the poorer sort bread is given. The number of children thus assembled are generally upwards of a thousand. At the bottom of the passage, there is a turn to the right and a flight of about twenty steps, four or five paces from the steps are a pair of folding doors that lead into the street. By some oversight the doors were not opened, the children in front, when they came to the bottom of the steps were buried alive by those who pressed from behind, in a short time the space from the doors to the top of the steps was one solid mass of human bodies. What made it still worse was that a parcel of beggars follow up the rear, to get a share of what is left, so that the poor little creatures in front had no chance to retreat from the danger, but were hurried into it, one on top of the other, until this large space from the top of the steps to the doors was completely crammed with upwards of 500 children. The confusion was beyond description, and before proper assistance

> would be rendered, upwards of one hundred had breathed their last breath.'

Private Wheeler relates that over a hundred children were killed and some 400 more injured. He concluded his letter:-

> 'At first the blame was thrown on the monks who are against the carnival, it was said it was done on purpose to try to do away with the carnival, but that cannot be possible, the fault must rest with the person whose duty it was to set the door open…'

CHURCHES ARE OTHER BUILDINGS where crowds gather, and the older the church the more likely it is that liturgical requirements will have dictated the size and shape of the building, with little thought given to the safety and comfort of the worshippers.

The Church of the Holy Sepulchre in Jerusalem is one of the oldest Christian churches and over the centuries it has suffered many vicissitudes (parts were burnt down in 1808 as a result, it is said, of a drunken monk accidently setting fire to some woodwork and then trying to douse the

flames with *aqua vitae* which he mistook for water). In 1834 it was the scene of appalling carnage, graphically described by a British traveller, Robert Curzon, who was in Jerusalem with some companions to witness the Greek Easter celebrations at the Church.

The ceremonies started in the evening of Good Friday and the pilgrims poured in. In Curzon's words '...the crowd was so great that many persons crawled over the heads of others, and some made pyramids of men by standing on each others shoulders... In consequence of the multitude of people and the quantities of lamps, the heat was excessive and a steam arose which prevented you seeing clearly across the church...' More worshippers arrived the following day to squeeze in as best they could. All were there to take part in the ceremony of the holy fire on Easter Day and, already in a state of extreme religious fervour, they were exhausted through standing all night in the heat and unbreathable air. The Greek Patriarch performed his brief rite and left the Church amid scenes of ever increasing frenzy. Three people, overcome by the heat and fatigue, fell from an upper gallery and were killed. A young woman died where she sat from heat, thirst and exhaustion.

Worse was to come.

> '... I saw a number of people lying one on another all about this part of the church, and as far as I could see towards the door. I made my way between them as well as I could till they were so thick that there was actually a great heap of bodies on which I trod. It then suddenly struck me they were all dead! I had not perceived this at first, for I thought they were only very much fatigued with the ceremonies and had lain down to rest themselves there; but when I came to so great a heap of bodies I looked down at them, and saw that sharp, hard appearance of the face which is never to be mistaken. Many of them were quite black with suffocation, and further on were others all bloody and covered with the brains and entrails of those who had been trodden to pieces by the crowd.'

To pile horror on horror, the panic stricken crowd then rushed the great door. The guards outside [Muslims] took fright and thought the erupting Christians were bent on attacking them. '... the confusion soon grew into a battle. The soldiers with their bayonets killed numbers of fainting wretches... Every one struggled to defend himself or to get away, and in the melee all who fell were immediately trampled to death by the rest.' Curzon himself barely escaped with his life. He states: 'Three hundred was the number reported to have been carried out of the gates to their burial-places that morning; two hundred more were badly wounded, many of whom probably died, for there were no physicians or surgeons to attend them, and it was supposed that others were buried in the courts and gardens of the city by their surviving friends; so that the precise number of those who perished was not known.'

As we have seen, pilgrims can be at some risk. An awkward little hazard came to light when Charles Greville joined the crowds at St. Peter's in Rome for the celebration of Easter in 1830. He noted in his diary:-

> 'To the Sistine Chapel for the ceremonies of Palm Sunday; we got into the body of the chapel, not without difficulty... It was only on a third attempt I could get there, for twice the Papal halbardiers thrust me back, and I find since it is lucky they did not do worse; for upon some occasion one of them knocked a cardinal's eye out, and when he found who he was, begged his pardon and said he had taken him for a bishop.'

GREVILLE WAS AMONG THOSE who disapproved of the ostentatious pomp of the Duke of Wellington's state funeral. His diary entry for 16th November 1852 reads:-

> 'I went yesterday to the lying in state of the Duke of Wellington; it was fine and well done, but too gaudy and theatrical... These public funerals are very disgusting *mea sententia*. On Saturday several people were killed and wounded at Chelsea...'

The Duke's embalmed body had been laid out in the Great Hall of Chelsea Hospital. So intense was the jostling of the grieving but undisciplined mob wishing to pay homage to their hero that the inevitable happened.

ALTHOUGH THIS IS NOT THE PLACE to consider the unspeakable records of the slave trade, it must not be forgotten that the cramped confines of a ship have always been open to the potentially fatal consequences of overcrowding. On 28th May 1796 two unscrupulous brothers were brought to trial at the Old Bailey on a charge of murder.

William and John Mitchell were owner and master of a 36 ton vessel which was lying at Jersey when she was commissioned to take 120 discharged soldiers back to England. After calling in at Guernsey a gale sprang up and the Mitchells drove their passengers pell-mell down into the ship's hold. The soldiers were already in a surly mood, having been denied permission to go ashore at Guernsey to procure sufficient drinking water, and one of them ('Colin alias Ezekiel Franklin' as the trial records name him) was shoved headlong down and severely battered. The hatch was then bolted down. The report of the trial continued:-

> 'In this small hold 120 people remained all night without any communication either of air or water, though they were constantly calling out to the captain for God's sake to bring them some relief. In this horrible state many of them became delirious and beat, bruised and stabbed one another. When the storm abated, the hatchway was opened and no less than 57 persons were found dead, among whom was Colin alias Ezekiel Franklin...'

The Mitchells were charged with the murder of Franklin but found not guilty because of doubt whether his death was caused 'in consequence of the confinement, or the blows given him by his companions, or of the rough useage he received from the owner and master of the vessel.' It does not appear that any action was taken against William and John Mitchell over the deaths of the other 56 passengers.

PANIC, of course, is a human emotion and cannot be legislated for. In the same year as the Sunderland tragedy, 1883, the Brooklyn Bridge in New York was opened. One week later, twelve people were trampled to death when a crowd of 20,000 panicked on rumours that the bridge was unsafe.

And panic was responsible for loss of life in Macclesfield in 1798. The Rev. Dr. Coke reported from there on 2 July of that year:-

> 'This evening I went into the pulpit, the chapel being so full that many could not get in. While the congregation was singing the last two lines of the second hymn, an old woman cried out, 'The roof is coming down.' One of our friends, knowing there was no manner of danger, in a whisper desired her to hold her tongue; but it was too late. Almost universal cries and shrieks took place. The people immediately rushed out. The court, which is but small, was instantly crowded. None, I think, were hurt in the chapel; but in the court, just on the outside of the door, six women and a boy of four years old were thrown down, and, alas! alas! were trod to death! It was in vain to cry out, 'Sit still, for there is no danger.' Nobody gave the least attention. I never was so distressed in my life; it being the most awful event I ever witnessed.'

By the late 1880s theatres, music halls and the like were subject to certain elementary controls. But inspection and licensing did not apply to private clubs and thus it was that the Hebrew Dramatic Club at Spitalfields in London was able to function in a totally unsuitable building. The stairs to the gallery were inside the main hall, from which the only exit was a single doorway leading to a vestibule that opened onto the street.

On 18th January 1887 an audience of some four hundred men, women and children were packed together in the gaslit hall to watch the drama being staged. The panic started when some young men in the gallery, trying to get a better view, pulled themselves up by grasping some gas piping. The pipe cracked, but although a quick witted man plugged the leak until the gas was turned off, it was by then too late. For a member of the audience, smelling the escaping gas, had called out 'Fire'. With the hall now in darkness, the terrified audience hurled themselves towards the exit, those in the gallery cramming the narrow stairway and merging with the stampeding crowd from the main hall. Sixteen people, mostly women and children, were crushed to death within the space of five minutes.

The inquest jury gave a verdict of accidental death, adding a rider:-

> 'That all places of this and similar kind, whether used as private clubs or other, should, to prevent a future recurrence of the calamity that had taken place, be placed under the immediate supervision of some public representative body.'

THE TRAVELLING PUBLIC, as we have noted in a previous chapter, is exposed to a variety of hazards, each one stemming from the particular mode of travel. A railway station should be a haven from such dangers, but here again much depends on the design of the station. Towards the end of the bank holiday in April 1892 the crowds who had been enjoying themselves on Hampstead Heath noticed an approaching rain cloud, decided to call it a day and rushed to the nearest station. The stairway down to the platforms at Hampstead Heath Station was soon jammed with a seething mass of passengers; somebody tripped and fell, but still the crowds came. Two adults and six children were killed. A passenger on a train arriving at the height of the crush described the scene in a letter to the *Hampstead and Highgate Express*:-

> 'A most painful sight met my gaze. The station seemed like a howling wilderness, shrieking, bustling, and cries of women and children made it a scene almost indescribable.'

The railway company came in for much criticism at the inquest ('... the arrangements made by the company were totally insufficient to cope with the increased traffic on public holidays...'). The public and press were outraged. A question was asked in Parliament and the President of the Board of Trade promised that improvements at Hampstead Heath Station would be in place before the next public holiday.

LORD CHESTERFIELD (the one who wrote all those tedious letters to his son) had no great opinion of the firework displays of his day. On 25th April 1749 he wrote to his friend Mr. Solomon Dayrolles about the galas planned to celebrate the peace treaty signed the previous year:-

> 'The next day come the fireworks, at which hundreds of people will certainly lose their lives or their limbs, from the tumbling of scaffolds, the fall of rockets, and other accidents inseparable from such crowds. In order to repair this loss to society, there will be a subscription masquerade on the Monday following, which, upon calculation, it is thought, will be the occasion of getting about the same number of people as were destroyed at the fire-works..'

In the spring of 1770 Marie Antoinette married the Dauphin of France, the future Louis XVI, and among the festivities held in Paris to celebrate this union was a great public spectacle in what is now the Place de la Concorde. There was a set piece fireworks display, the grand houses were illuminated and the trees in the square were hung with lamps. A fair was to be held in the Rue Royale.

An official report described what happened. After the fireworks some 300,000 men, women and children left their vantage points and converged on the Rue Royale for the fair, some in carriages, most on foot. They crossed with other crowds making their way to see the illuminations. Many stumbled over the street gutters and were trodden underfoot; others tripped over the fallen bodies and added to them. Panic set in. Carriages were overturned and so many tried to climb over them that they collapsed under the weight and horses too were suffocated. The dead were carried along upright by the crowds, only falling down when no longer supported by the press of people. Scaffolding erected over the Seine to provide a good view of the fireworks proved inadequate for the crowds assembled on it and gave way. Hundreds were drowned.

All in all nearly 1,000 were said to have lost their lives that evening. The following morning the Dauphin wrote from Versailles to the authorities in Paris:-

> 'I have learned of the disaster which came upon Paris on my account. I am deeply distressed. I have brought the sum which the King sends me every month for my private expenses. It is all I have to give. I send it to you. Give aid to the most unfortunate.'

There is a sad little footnote to this tragedy. One hundred and thirty-two identified victims were laid out in a small enclosure in the graveyard of the Madeleine awaiting burial. Into this same enclosure, 23 years later, the body and guillotined head of Marie Antoinette were tossed contemptuously and lay there until the grave digger, inappropriately named Joly, got round to doing his job.

HAPPILY THERE IS ONE SPECTACLE that has disappeared down the plughole of history and that is the public execution. The last person to be beheaded in England was Simon Fraser, Lord Lovat. He was a fervent Jacobite, was found guilty of treason for his part in the 1745 rebellion and in April 1747 was beheaded on Tower Hill in London. Spectators crowded the roof and battlements of the Tower and the stands that had been built to accomodate them. Lord Lovat himself expressed surprise at the numbers: 'God save us! Why should there be such a bustle about taking off an old grey head...' He was 80 years old at the time, and so great was the bustle that one of the stands collapsed under the weight of the thousand-odd spectators on it, crushing to death some twenty people.

The public hanging of criminals continued into the next century on a misguided interpretation of the principle that justice must not only be done, but must be *seen* to be done. (Edinburgh even had a kind of hospitality tent in the shape of a pub called The Last Drop which stood opposite the gallows in the Grassmarket, just below the castle.)

Such events attracted excitable mobs and feelings ran high, as they did on 19th February 1807. On that day in London a number of convicted murderers were to be hanged, and 40,000 avid voyeurs turned up to watch this public execution. Two piemen with their fast food wares joined the crowd in the hope of turning an honest penny. Just before the condemned men mounted the scaffold a great cry of 'Murder! murder!' went up and the crowd became increasingly excited about the approaching spectacle. One of the piemen had his basket knocked from his hand. He scrabbled on the ground to retrieve his spilled pies,

> *But those behind cried 'Forward!'*
> *And those before cried 'Back!'*

In the crush that followed, 32 spectators were trampled upon and suffocated.

# CHAPTER 10

# THEY SHOULD HAVE KNOWN BETTER

THERE IS ALWAYS GREAT SATISFACTION to be gained at the spectacle of the experts being defeated by their own speciality. Unchristian, no doubt, but this *schadenfreude* is all too human and few of us are highminded enough to feel undiluted pity for the greengrocer who slips up on his own banana skin.

Take an incident which introduced a moment of farce into the 1605 Gunpowder Plot. The story is well documented. The intention of the conspirators was to blow up the Houses of Parliament while King James I was there for the State Opening, to kidnap the heir to the throne and to join an uprising of disaffected citizens in the Midlands which they believed the death of the King would set in train. An anonymous tip-off to the Government led to the discovery of Guy Fawkes and his gunpowder in the cellars beneath the House of Lords. Fawkes was arrested and the other plotters hastily left London. Some of them, including Robert Catesby, the instigator of the whole plot, reached Holbeach House in Staffordshire, after hard riding in the rain, carrying with them two bags of, by then, very wet gunpowder.

Amazingly, the Gunpowder Plotters spread out the contents of one of the bags of powder to dry before a crackling fire, the other bag lying close by. The inevitable happened and the loose powder exploded. Catesby was one of those seriously injured. The author of *A True and Perfect Relation of the Whole Proceedings against the late most Barbarous Traitors*, Imprinted (1606) at London by Robt. Barker, Printer to the King's Most Excellent Majesty, evidently felt as we do, but commented self-righteously: '*In quo peccemus, in eodem plectimur*'. ('In the manner in which we sin, in the same manner shall we be punished.')

MOST OF US LEARNED about the Gunpowder Plot in our schooldays. Those of us who were also taught chemistry (or stinks, as uncouth schoolboys and graceless schoolgirls generally called it) would have learned that:-

*Poor Willie's gone and left us now,*
*His face we'll see no more,*
*For what he thought was* $H_2O$,
*Was* $H_2SO_4$.

A close relative of Willie must have been the Professor of Chemistry at Grenoble University in France. In 1880 he gave a lecture on mercury and its salts. He had on the bench in front of him the customary glass of water to refresh his throat during the lecture. He also had a beaker containing a solution of a mercury salt. In his enthusiasm for his subject he picked up the beaker by mistake and swallowed the contents. The unfortunate Professor died immediately.

Chemists do rather seem prone to such accidents and the layman may wonder whether they know as much about the substances they handle as they should do. Had he lived, the poor laboratory assistant in the following incident might well have opted for another career. It seems he collapsed while making hydrogen cyanide in a fume cupboard. His colleagues, with the best of intentions, gave him an antidote made by pouring copper sulphate crystals into a beaker, adding water and stirring vigorously. The lab assistant died. An autopsy was carried out by a pathologist who expressed the view that he had died of copper sulphate poisoning. To compound this unfortunate mishap, a sample of the stomach contents was given to the police for formal analysis but they managed to lose it.

Reverting for a moment to that *enfant terrible* of chemistry, Willie's own father did not seem to know all he should have done about appropriate antidotes for poisoning:-

*Willie finding life a bore*
*Drank some* $H_2SO_4$
*But Willie's father, an MD*
*Gave him* $Na_2CO_3$
*And now he's neutralised, it's true*
*But he's full of* $CO_2$.

EVEN THE FAMOUS have their lapses. The importance of oxygen in the combustion process was established in the 18th century by the renowned chemists Lavoisier and Berthollet. When therefore Berthollet discovered potassium chlorate in his laboratories at Essonne in France he realised its potential for releasing oxygen and sought a use for his discovery. Replacing potassium nitrate in gunpowder sprang to his fertile mind. In 1788, he demonstrated the preparation of his new black powder in a crushing mill in the presence of many invited guests. The operation was nearly completed when a stupendous explosion took place, killing an assistant and the daughter of the French Government Commissioner of Explosives.

The skilled and experienced French chemist Pierre Louis Dulong (1785–1838) lost an eye and three fingers in explosives research.

*The many perils that environ*
*The man who meddles with a siren*
*Are naught beside the ones that he*
*Invites, who flirts with TNT.*

And the great Michael Faraday had a few mishaps too. He wrote to a friend in 1813 about some explosions resulting from his work with unstable chemicals:-

'Of these the most terrible was when I was holding between my thumb and finger a small tube containing about 7 1/2 grains of it. My face was within 12 inches of the tube but fortunately I had on a glass mask. It exploded by the heat of a small piece of cement that touched the glass above half an inch from the substance and on the outside. The expansion was so rapid as to blow my hand open, tear off part of one nail and has made my fingers so sore that I cannot yet use them easily.'

Ten years later he was liquefying gases and still writing to friends about his self-inflicted wounds:-

'I met with another explosion on Saturday evening, which has again laid up my eyes. It was from one of my tubes, and was so powerful as to drive the pieces of glass like pistol-shot through a window. However, I am getting better and expect to see as well as ever in a few days. My eyes were filled with glass at first.'

IN MORE MODERN TIMES another explosion had altogether happier results. In his final year at school young Ted used to help out at the local university, where he assisted the Professor of Chemistry in setting up experiments for demonstration to the students. One day he had to set up an experiment to electrolyse water to hydrogen and oxygen. The gases came off the electrodes separately and each was passed by a pipe under a soap solution. Each of the gases could be collected separately or if the two pipes were put together, bubbles of the mixed gas would arise. The Professor would demonstrate the properties of each gas by putting a lighted taper into a gas bubble; an oxygen bubble would make the taper burn fiercely and the hydrogen bubble would give a small bang. The apparatus was set up but the Professor was delayed and Ted, wishing to demonstrate the gases himself, switched on the electrical current. Many bubbles frothed over the dish of soap solution. He plunged the taper into the bubbles and there was an explosion which blew out the windows of the laboratory.

Doubtless remorse at his youthful arrogance prodded Ted into becoming the greatly respected member of the Institution of Chemical Engineers and indefatigable worker on loss prevention that he is today.

IN THE 13TH CENTURY they dealt more robustly with the professional who made a hash of things. In 'The sea law of Olearon', preserved in *The Oak Book of Southampton*, mention is made of the likely consequences:-

'Because great loss is often caused by pilots, for that he [the pilot] agrees to take a ship into a harbour, and steers her badly, whereby many a master and many a merchant is impoverished, it is our wish that if he undertake to steer a ship into any harbour to safety, and he steers her badly and the ship is lost through his not knowing the entrance into the precincts of the harbour, the statute ordains that he lose his right hand and his left eye, because he has treasonably led them. And this is the judgement in this case.'

Given the primitive navigational aids of the time and the punishment for failure, we cannot imagine there were many applicants for a job as pilot.

One of the primitive aids was, of course, the magnetic compass, whose properties had been known for centuries when the *Great Britain* was launched in 1843. She was the largest vessel that had then been built but she had a steel hull and, astonishingly, nobody seems to have appreciated the effect such a hull would have on the compass. Nevertheless, she made four successful trans-Atlantic runs with a compass showing out-of-true readings but which, by good fortune, were not of importance on those crossings. It was only when she hit really bad weather off the Northern Ireland coast in 1846 that her experienced master was forced to rely heavily on a compass whose readings were distorted by the proximity of all that steel, with the result that his ship ran aground at Dundrum Bay. The problem was belatedly identified and when the *Great Britain* was refloated an attempt was made to distance the compass from the hull by placing it at the masthead and viewing through a periscope. This was hardly a practical solution and eventually two compensating metal balls were positioned next to the compass to ensure true readings.

IT IS SAID that the second man (the first was Benjamin Franklin) to charge a Leyden jar from lightning was killed by the experiment. This happened in St. Petersburg, but from Canada comes the report of an early researcher into cardiac rhythms. George Mines was 28 years old in 1914. In his laboratory at McGill University in Montreal he made a small device capable of giving small, precisely regulated electrical impulses to the heart. When he decided it was time to begin work with human beings he chose the most readily available experimental subject — himself. At about 8 o'clock one evening a caretaker, thinking it was unusually quiet in the laboratory, entered the room. Mines was lying under the laboratory bench surrounded by twisted electrical equipment. A broken mechanism was attached to his chest over the heart and a piece of apparatus nearby was still recording the faltering heartbeat. He died without recovering consciousness.

SIR GEORGE AIRY, a man of impressive qualifications, was appointed Astronomer Royal in 1835. Earlier, as a Fellow of Trinity College, Cambridge, he had set about 'weighing the Earth' by measuring gravity from the bottom of two deep mines. In the first one the barrel containing equipment packed in straw caught fire as it was being raised, probably ignited by a candle tossed in by resentful miners. The experiment in the second mine was halted by the fall of a rock, described by Airy as 'many times the size of Westminster Abbey.'

Never burdened by humility, George Airy confidently predicted that the

Crystal Palace would blow down (it didn't) and that the Atlantic telegraph cable would not withstand the pressure of 'so great a depth' (it did). In view of such ill-founded pronouncements it is astonishing that Thomas Bouch, the engineer who built the first Tay Bridge, should have trusted Airy to check the mathematics of wind pressure on his proposed railway bridge. The Astronomer Royal comfortably asserted that 'The greatest wind pressure that a plane surface like that of a bridge will be subjected to in its whole extent is 10 lbs per square foot.' Just how misplaced was Bouch's trust in Airy became a matter of history when the Tay Bridge was blown down in 1879.

WE ARE UNABLE to fathom the reasoning of those who do not use the safety equipment recommended and provided. Hard hats, dust masks, life jackets, safety harnesses, all are dispensed with at times by those who think they know better. They delude themselves. A couple of examples from the records of the Health and Safety Executive will suffice.

One concerns a fitter, using a grinding wheel, who had twice been admonished for not wearing his safety spectacles. He was thereafter seen to be wearing them regularly. But then he had to seek medical treatment for a fragment of metal in his eye. An investigation revealed that the fitter had pigheadedly removed the glass from his safety spectacles. Another investigation, this time into a fatality in a tank being cleaned of oil sludge, disclosed the astonishing fact that the contractor's men had removed the plastic visors of the breathing apparatus, which are vital to such work in a fume-filled enclosed space. An additional folly also came to light; one of the men had lit a cigarette, causing the fire which led to the death of another man.

More safety conscious than the employees just mentioned was the chief fire officer at a chemical works but even he fell below his own high standards. He was justifiably proud of his fire crew and frequently 'complained' that he could never get to a fire before the crew had put it out. One Saturday, around lunch time, he was telephoned at home with the news that there was a fire at the works. He dropped everything and hastened to the works, to be thwarted once again — the fire had been extinguished. After checking that the fire was truly out and determining its cause, he returned home to find his kitchen on fire. He had left a chip pan on a lighted cooker.

SUCH FORGETFULNESS may be excusable in an emergency. Not so when Richard Trevithick took his steam road carriage for its first trial run on Christmas Eve, 1801. The great Cornish engineer knew all about steam power from his work on the high pressure engines he developed for the tin mines of Cornwall, and he spent a year constructing his road carriage.

For the trial he chose a hill with a 1 in 20 gradient. The vehicle broke down after travelling only a few hundred yards, and Trevithick and his companions were obliged to move it off the road into an outhouse. They then took themselves off to a nearby inn where, according to one who was present, they 'comforted their Hearts with a Roast Goose & proper drinks.' But Trevithick had committed the driver's unpardonable sin; he left his engine running. In other words, he 'forgot to extinguish the Fire that evaporated the water and then heating the Boiler red hot, communicated fire to the wooden machinery and everything capable of burning was consumed.' Even the building into which the steam carriage had been pushed was burned to the ground.

GUNSMITHS can be expected to show great respect for the items they handle and usually they do. However, the following accident is the more inexplicable because the gunsmith had been warned immediately before that the gun was loaded. It belonged to Hugh Miller, a great figure in the world of Victorian science who had the misfortune to suffer from a disease of the brain which manifested itself as a paranoid fear of being attacked. He took to carrying a loaded gun and one morning in 1856 he was found shot dead through the chest, his gun close by in the bath. The gun was rusted from lying overnight in the bath water and was taken to a gunsmith, Thomas Leslie. It was handed to him with the warning, 'Mind, it is loaded.' Leslie examined the rusty safety catch, held the gun to his eye and lifted the hammer to count the remaining bullets. The gun went off through his eye and blew his brains out.

MANAGERIAL IGNORANCE was pleaded as a mitigating circumstance in 1887 when explosions at a chemical company outside Manchester were investigated. The company produced picric acid for the dyeing industry and at the time of the incident was understaffed, many employees being away for the celebrations of Queen Victoria's Golden Jubilee. On the 22nd June there were two explosions in rapid succession, powerful enough to blast a baby from her cradle in a nearby cottage and to be heard over 20 miles away. Some of the work people had to jump from storey to storey as the staircase had been destroyed in the explosions. A fireman in the service of the company was killed when a wall fell on him.

At the subsequent inquest the company blamed the disaster on a labourer smoking a pipe and even denied they knew that they were producing chemicals of an explosive nature — this despite the fact that some fifteen years earlier chemist Sprengel had shown the explosive power of picric acid. The verdict put the blame on the employee for smoking. The company was only mildly rebuked for being ignorant of the chemical properties of their products.

There is said to be a tombstone in the churchyard at Collingbourne Ducis in Wiltshire, bearing the following epitaph:-

*Blown upwards out of sight,*
*He sought the leak by candle light.*

The burial place of the simpleton at the heart of the following accident is not known, but his final exploit, which caused an explosion, lives on in the 1834 Select Committee report on the state and management of lighthouses, floating lights, buoys and beacons. It seems he attempted, by the light of a candle, to repair a faulty gasometer in the gas-lit pierhead light tower at Holyhead.

AN EXPLOSION on 5th July 1880 was triggered off by a foreman lighting a match in circumstances when experience and common sense should have promoted caution. It happened in the Tottenham Court Road area of London, where a new section of 36 inch gas main had been laid and connected by a valve to an existing main. The new section was plugged off and left for a few days while another section was built. Eventually the plugged-off section was checked for pressure by a pressure gauge connected to a half-inch standpipe. No pressure was indicated but to check for the presence of gas, the foreman applied a match to the open end of the standpipe. Nothing came out but about one minute later there was a rumbling noise followed immediately by an explosion in nearby Percy Street. There were then two other explosions in Percy Street and four in Charlotte Street close by. The final explosion at the valve with the main was the most severe, with sections of the pipe being blown out of the ground.

The two foremen at the site, with 45 combined years experience in the gas industry, had not learnt that a confined explosion could occur and expressed astonishment at the incident. It was from this incident that chemists learned flame velocities could reach 100 yards per second.

It was the action of an experienced miner which precipitated an explosion in 1869. It happened in the city of Cleveland in the USA, where the authorities decided to extend the water tunnel some 6,600 ft out under Lake Erie. Local workmen were taken on to sink the shaft, and the foreman employed had learned his skills in English coal mines. The work started in August 1869. By the time the shaft had been sunk 63 ft with only 4 ft remaining for tunnel grade to be reached, the miners struck gas. The foreman went down with one of the workers to investigate. When the bucket was half-way down, this experienced coal miner, quite unaccountably, struck a match. He and his companion in the bucket were killed instantly. The explosion and fire also severely burned two men at the head of the shaft.

There was a whimsical finale. One man standing on a plank laid across the top of the shaft was blown 12 ft into the air, along with the plank. When he descended, the plank had fallen back into place a fraction of a second earlier and was there to receive him. He was the only one around the shaft to be unhurt.

THE TREATY SIGNED AT AIX-LA-CHAPELLE in October 1748 brought to an end the War of the Austrian Succession, but the peace was not publicly celebrated until the spring of 1749. Since fireworks were the order of the day, the pyrotechnicians came into their own. And a pretty poor job they made of it too. In Paris, according to a letter Horace Walpole wrote on 3rd May, '... there were forty killed and near three hundred wounded, by a dispute between the French and Italians in the management, who, quarrelling for precedence in lighting the fires, both lighted at once and blew up the whole.'

Such squabbling over procedure and precedence was hardly conducive to that European harmony the French and Italian pyrotechnicians were employed to celebrate. But there was no EU directive to regulate matters, and at the London celebrations a remarkably similar falling out of experts led to disaster.

A grandiose structure, 410 ft long by 114 ft high, designed by Cavalieri Servandoni, had been erected in Green Park, from which 11,000 fireworks were to be fired to the *Music for the Royal Fireworks*, composed by Handel for the occasion. Gaetano Ruggieri, renowned in firework circles throughout Europe, supplied the fireworks for this extravaganza and came over from Italy with them. He found on arrival that he was to share responsibility for firing the display with several others, among them Captain Thomas Desaguliers, the firemaster, Charles Frederick, Comptroller of the Woolwich Depot, the Royal Train of Artillery, and two local pyrotechnicians. No single one of them was in charge; each thought he had full authority and each issued his own orders. Inside Servandoni's 'machine' the English and Italians fell to quarrelling about the best method of igniting the fireworks and, with their attention thus diverted, failed to prevent an explosion in the North Pavilion of the structure. It burned to the ground and almost brought down the whole of Servandoni's grand setting. The destruction of part of his magnificent edifice so enraged Servandoni that he drew his sword on the first pyrotechnician he came across, who happened to be Charles Frederick. He was disarmed, arrested and consigned to the Tower of London until he apologised for his part in the affray.

Horace Walpole thought the whole spectacle 'pitiful and ill-conducted' and twenty-five tons of fireworks were left unexploded. They were acquired by the Duke of Richmond, who gave a private and hugely successful display some two weeks later.

THE STAGE WAS ALL SET for the explosion that occurred in 1911 at a government powder magazine in Costa Rica. A single building was divided into two rooms separated by a brick wall. In one room was stored some 23,000 lbs of explosives, while in the other three or four soldiers lived, slept and cooked. After the explosion which killed the four soldiers and a child and injured several other children, witnesses revealed some astonishing facts. Guards on the magazine had been seen smoking, and one witness stated that on the day before the accident he had seen one of the soldiers sitting on the roof of the building whittling and smoking cigarettes. An agent sent some years previously to destroy old dynamite (so old that it was in a dangerous condition) said he had found great difficulty in persuading the officer in charge to throw away his cigarette at the door of the powder room. He added that the floor was a quarter of an inch deep in loose black powder, and only by refusing to enter could he induce the officer to remove his hobnailed boots.

The owner of a gunpowder mill in Massachussetts in the United States seems never to have asked the important question Why? His mill blew up in 1820, killing four men. In 1821 it blew up twice, causing three deaths. Shrugging his shoulders, he relocated his mill a short distance away; yet more explosions followed, the most devastating being in 1826 and 1830. The product, known as Boston Gunpowder, was said to have been of the highest quality but we do wonder how they ever managed to make enough of it to find out.

BUT FOR TOTAL DISREGARD of past accidents and failure to take appropriate action, the Earl of Berkeley surely cannot be faulted. As reported on 21st September 1789:-

> 'A gamekeeper belonging to the Earl of Berkeley, on setting his Lordship's spring-gun in the paddock, was shot dead on the spot. What is remarkable is that in the course of four years, by the same gun, three gamekeepers have lost their lives, and one gentleman shockingly wounded in fixing the wire to the ground.'

This Earl, who married a butcher's daughter (and later forged an entry in the Marriage Register to please her), led a life of some vigour. He shot and killed at least two highwaymen who were trying to rob him and almost came to a sticky end himself when attacked by one of his own deer in the rutting season. But he must be forgiven, if only because he introduced his friend Edward Jenner to the Royal Family, with the happy result that vaccination against smallpox was examined by a Parliamentary Committee and could be said to have Arrived.

# CHAPTER 11

# CLOSE CONFINEMENT

MANY HAZARDS OF THE PAST are still hazards today, and all too frequently no lessons have been learned from the accidents they have caused. Entry into a restricted space is one of these dangers. In early times mining and the sinking of wells were the common occasions of what is technically known as confined space entry, and in AD 79, when he wrote his *Natural History*, Pliny the Elder stressed the importance of ventilation:-

'...that besides the principal shaft, it was the practice to sink vent holes on each side of the well, both right and left, in order to receive and carry off the noxious exhalations. Independently of these evils, the air becomes heavier, from the greater depth merely of the excavation, an inconvenience which is remedied by keeping up a continual circulation with ventilators of linen cloth.'

The botanist Charles Linnaeus after visiting a mine in 1734 wrote:-

'The sulphurous smoke poisons the air and kills everything growing, and fills the cavities of the mine with evil fumes, dust and heat. Here 1,200 men labour shut off from light of the sun, slaves under the metal, less men than beasts, surrounded by soot and darkness. Fear of being crushed under falling rock never leaves them for one moment. These damnati work naked to the waist and have before their mouths a woollen cloth to prevent them from breathing too much smoke and dust. They cannot take a breath of pure air ....'

AS ANY WELL INFORMED CANARY will tell you, working in a confined space is a highly dangerous occupation and the importance of testing the atmosphere in a vessel, mine, well or any other enclosed area has been emphasised on many occasions but, sadly, is frequently ignored.

There is a gravestone in the churchyard of St. Cuthbert's Church at Marton-in-Cleveland, near Middlesbrough. It reads:-

'Erected in memory of Robert Armstrong aged 28, James Ingledew aged 39 and Joseph Fenison aged 27 years who

> unfortunately lost their lives on the 11th February 1812 by venturing into a well at Marton when it was filled with carbonic acid gas or fixed air. From this unhappy accident let others take warning not to venture into wells without first trying whether a candle will burn in them, if the candle burns to the bottom they may enter with safety. If it goes out, human life cannot be supported.'

The background of this sad epitaph is on record in the *Annual Register* of 1812. Two wooden plates had fallen into a well, and Robert Armstrong was lowered in a bucket by two friends to recover them. When some way down, he collapsed and fell out of the bucket. The others shouted to him but got no reply, for Robert was unconscious at the bottom of the 60 ft well which, however, contained only 2 ft of water. James Ingledew was then lowered in the bucket but also collapsed and fell out when half way down. Joseph Fenison thereupon went down with the help of a ladder. He fell off it into the bottom of the well. Last came William Hardwick, who had the good sense to tie a rope round his waist before descending. He too collapsed and was hauled out barely living, but happily he survived. The bodies of the three dead men were eventually recovered by means of a grappling iron.

William Hardwick used a rope to tie around his waist, but here is a heroine who tied herself to a rope with her own hair. On the 9th April 1788 it was reported from Noyon in France:-

> 'On 1st of this month, about eleven o'clock at night, four men were suffocated by the mephitic vapours occasioned by the opening of a common sewer, into which they unluckily fell. At that unseasonable hour it would have been difficult to get proper assistance and the necessary succours from the Humane Society; they must have inevitably perished, had not a young woman of seventeen, servant to the family, begged to be let down in order to save them. This woman, whose name is Catherine Vassent, went down seven different times, and succeeded so far as to save two of them pretty easily; but in tying the third to a rope, which was let down to her for the purpose, she felt her breath failing her, and was very ill. In this terrible situation she had the presence of mind to tie herself by the hair to the rope. and one may judge of the surprise and alarm of those who drew up the expiring girl close tied by the side of a dying man. The moment she recovered her spirits, she insisted on going down for the fourth, but it was to no purpose, he was drawn up dead.'

THE MARTON EPITAPH given above stressed the importance of candles in testing the atmosphere but they had long been used by miners to ignite firedamp at the coal face. A man called the 'fireman' had the unenviable task of testing the air in mines for flammable gas, and a 17th century writer gave his job description thus:-

> 'The ordinary way in which the hurt of it [the gas] is prevented, is by a person that enters, before the Workmen, who being covered by wet cloth, when he comes near the Coal-wall, where the fire is feared, he creepeth on his belly, with a large pole before him, with a lighted candle on the end thereof, with whose flame the Wild-fire meeting, breaketh with violence, and running along the roof goeth out with a noise, at the mouth of a Sink, the person that gave fire, having escaped by creeping on the ground, and keeping his face close to it, till it be overpassed which is in a moment.'

Only when any gas present had been ignited was the coal face considered a safe place to work in. In at least one mine a happy, if drunken, chance introduced the workmen to a slightly different method of clearing the atmosphere. The 17th century biographer, John Aubrey, described the incident:-

> 'Sir Paul Neale sayd, that in the Bishoprick of Durham is a Coalery, which by reason of the dampes there did so frequently kill the Workmen (sometimes three or four in a month) that he could make little or nothing out of it. It happened one time, that the workmen being merry with drink fell to play with fire-brands, and to throwe live-coales at one another by the head of the Pitt, where they usually have fires. It fortuned that a fire-brand fell into the bottome of the Pitt; whereat there proceeded such as noise as if it had been a Gun; they liking the Sport, threw down more fire-brands and there followed the like noise, for severall times, and at length it ceased. They went to work after, and were free from Damps, so having by good chance found out this Experiment, they doe now every morning throw down some Coales, and they work as securely as in any other Mines.'

The word damp, incidently, comes from the German *dampf*, meaning vapour, and the earliest recorded accident caused by firedamp took place in a pit at Gateshead. The register of St. Mary's Church there records, under the date 14th October 1621, the burial of 'Richard Backas, burn'd in a pit.'

Fire was not in all cases the answer to atmospheric problems in mines, and in at least one pit tragedy it was the causal factor. This was at Madely in Shropshire where, in 1810, there was a fire at the bottom of the coalmine, 240 yards below the surface of the earth. All those underground were brought up unharmed. Sadly, though:-

> 'Four persons went down the pit next morning to see what state it was in, and what was best to do, but with the sulphor becoming too powerful for the air, all four were suffocated.'

ANIMALS WERE LATER FOUND to be of great value in testing the atmosphere in confined spaces. (Around 1704 a certain Mr. W. Derham experimented with a sparrow and a Great Titmouse in 'one of Mr.Hawksbee's Compressing engines.') Canaries, in particular, have been used extensively in mines, partly because they can be heard chirping but mainly because they fall off their perch in concentrations of carbon monoxide below that which affects men.

THE USE OF CANARIES is, of course, common knowledge. It is less well known that white mice were also used up to the 1960s to test the atmosphere in enclosed vessels at some chemical plants. The white mice were kept by the works fire station and issued to the plant on completion of the appropriate form. The white mouse had to be inside the vessel for one hour before it was considered safe for a workman to enter; the mouse also had to stay in the vessel while the work was being carried out. A fresh mouse had to be brought in every four hours, but despite this the little creature had an undignified time of it.

Operatives wear dust masks but it is difficult to afford this protection to mice, who can breathe only through the nose, having no mouth-breathing back-up. In one plant, if a mouse was seen to be in difficulty it was warmed up on a radiator and rushed back to the fire station before it died, since an inquiry — or rather, an inquest — had to be held if a mouse died on the plant premises. On one occasion an alkyd resin reactor was prepared for entry by maintenance workers and a white mouse was placed inside for the customary hour. The foreman checked the mouse half an hour later and found it dead. Numerous tests were carried out on the reactor atmosphere to discover the gas which had caused the death but none was found. Because it was vital to the safety of the workmen to find the cause of the mouse's death, an autopsy was carried out on the little body at the local hospital; the mouse was found to have died from natural causes.

Meanwhile, the maintenance work for which the white mouse had been taken on strength had been held up for 24 hours while the reason for its death was being established.

RESEARCH ON HUMAN RESPIRATION was carried out in 1872 by the Rev. George Garrett Pasha (as well as being a cleric in the Church of England he was an officer in the Turkish Navy):-

> 'During these investigations [he] allowed himself to be enclosed in an hermetically sealed chamber, which contained besides himself a bird, rabbit, guinea pig and a lighted candle. He took samples of the air in the chamber every fifteen minutes, as well as working out mathematical problems [to see how his consciousness was being affected]. He remained under observation until he lost consciousness, which was not until the animals had passed out and the candle extinguished.'

The 20 year old Garrett used the results of this work to design the world's first self-contained breathing apparatus, using caustic potash to scrub out carbon dioxide from exhaled air, an invention that would finally appear five years later as the vital element of what he named a Pneumatophore.

The concern for carbon dioxide was also shown in 1880 when a Mr. Fleuss was experimenting with a diving suit at Fishbourne Creek, Isle of Wight. The apparatus, known as an oxygen rebreather, consisted of a totally closed circuit system with soda lime to absorb the exhaled carbon dioxide. Mr. Fleuss was aware of the importance of removing the carbon dioxide but the lack of oxygen led to his loss of consciousness. Fortunately, he was rescued and later developed a self-contained breathing apparatus using compressed air carried in a knapsack. This equipment was used by a diver in 1880 to repair a valve in Brunel's tunnel under the River Severn when it was flooded during construction work.

AN ACCIDENT SIMILAR to the one at Marton happened in France in 1786. It was reported in the *New Annual Register* for that year:-

'Paris, October 3. They write from Besiers that some workmen employed in digging a well at Antignac, a village three miles from thence, got to a depth of about six toises [sic] the third of last month, when, observing water to rise, they redoubled their activity, and were presently astonished by a most violent subterraneous explosion. Having recovered from their surprize, they again approached the pit, at the bottom of which they perceived one of their comrades, to whom they called, but received no answer. One of his brothers being apprehensive for his safety, descended in a bucket in order to yield him assistance; but this man showed no sign of life after he had reached the bottom. He was followed by a third, who experienced the same fate. A fourth had the courage to descend, his companions taking the precaution of fastening a rope to him; and following him with the eye, as he was gently lowered, they soon perceived his head to droop, and his whole frame to be violently agitated. Being immediately drawn up, he continued without motion for two hours.

'Recourse was now had to experiments which ought to have been first adopted. They let down a cock in a bucket, and on being drawn up it was found on the point of expiring, with its feathers burnt. The same was done with a cat, which was almost dead when drawn up. By means of hooks and other implements the three persons were raised out of the pit, being quite lifeless, and all their skin appearing to be calcined. The letters further say that the subterraneous noise still continues and that chemists are endeavouring to discover the cause of the explosion, and the vaporous gas which has proved so fatal in its effect. It is added that vitrified matter has been taken from the pit, which, it is supposed, must have been in a state of fusion.'

It is of interest that this 18th century report also emphasised the wisdom of first testing an atmosphere before sending men into confined spaces.

AT THE END of the nineteenth century Professor Haldane arrived on the scene and interested himself in the physiology of respiration. He investigated many accidents, one of which had taken place in 1895 in a 27 ft deep well at the East Ham Sewage Works in London.

Five men had been killed. The first victim experienced breathing difficulties and tried to climb the ladder to the surface; he was overcome by fumes, fell back into the sewage below and drowned. His three colleagues went down to rescue him but were also overpowered; two of them plunged to the bottom and died instantly, but the third was caught up in some staging as he fell. The engineer in charge descended the ladder to help the man on the staging but he too was overcome and fell into the sewage.

Haldane identified the gas responsible ('sulphuretted hydrogen'), and his investigation led to the adoption of certain precautions such as safety ropes for workers in confined spaces and the ventilation of sewers before entry. Another accident he looked into was an explosion the following year at a coal mine in South Wales. He found that of the 57 fatalities only five men had been killed by the force of the explosion; the remaining 52 had died from carbon monoxide poisoning.

TUNNELLING OF ITS VERY NATURE presents no end of problems and there is a long history of underground accidents. One of the most ludicrous must surely be a startling incident in New York in 1905. Tunnel workers there were known as sandhogs. Eight of them were tunnelling, in a shield, under the East River, when compressed air blew a hole through the tunnel crown, at that point only 5 ft below the river bed. One of the sandhogs tried to stop the leak with a bale of hay but was blown upwards into the hole. He was briefly stuck in the silt with only his legs dangling below. The air pressure then shot him up through the silt and propelled him ever upwards through 15 ft of water to the surface of the river. He was rescued unharmed by some understandably astonished longshoremen.

# CHAPTER 12

# THE BEST LAID SCHEMES

OUR MORE REMOTE ANCESTORS took the precaution of consulting an oracle when about to embark on a great project. The one at Delphi was a favourite. Peering at the entrails of dead animals was also popular. Sometimes a soothsayer would issue bulletins unbidden, and our own Mother Shipton, who died in 1561, is famous for her output of predictions. Relevant to our purpose are the road accidents she foresaw:-

*'Carriages without horses shall go*
*And accidents fill the world with woe.'*†

But for every Mother Shipton there is a Peter of Pomfret, who prophesied in 1213 that King John's reign would come to an end on Ascension Day of that year. The King did not take too kindly to this, and Shakespeare has him respond:-

*'Hubert, away with him; imprison him:*
*And on that day at noon, whereon, he says,*
*I shall yield up my crown, let him be hang'd.'*

Ascension Day passed without mishap to the King, and Peter was hanged at Wareham in Dorset on some earthworks ever since known as The Bloody Bank.

IT IS WORTH NOTING that when future disaster is prophesied accompanying preventative measures are rarely foretold, unless it be apocalyptic directions to turn away from sin to deflect the wrath of the Almighty. Mother Shipton did not suggest the carrying of red flags in front of her carriages-without-horses. Always bearing in mind Murphy's dictum that no matter what happens there is always someone who knew it would, recourse to a clairvoyant is clearly not a reliable way to go about ordering matters of safety. Instead, we look to more logical and scientific methods of

†Roger Bacon in the 13th Century had foreseen motor cars but not the accidents. He wrote 'Likewise carriages can be built which are drawn by no animal but travel with incredible power.' Just how incredible the power we know only too well.

foreseeing a hazard so that steps can be taken to remove it, and we have, over the years, developed all manner of safety devices and procedures, drills and training. Nevertheless, however carefully we scheme to identify and eliminate dangers, always lurking in the background are human negligence, ineptitude, bravado, inbuilt flaws, over-enthusiasm, misunderstandings and downright bloody-mindedness to frustrate the best endeavours of safety-conscious individuals.

The builders of the River Mersey railway tunnel in the nineteenth century deserve our sympathy. Every care was taken to ensure the safety of the workmen, and rigorous inspection was the order of the day. An engineer regularly visited the tunnellers at work and bored a hole with a 15 ft drill into the roof of the tunnel, without once piercing the riverbed, to reassure the men that they were well below the river and in no danger of a sudden inrush of water. But a fine safety record was marred by the irresponsibility of eight young miners descending in a lift cage at the start of their shift. They indulged in what the press referred to as 'sky-larking', upsetting a petroleum lamp and setting the cage on fire; two of the men jumped down the shaft and were killed on the spot, while the others were badly burnt.

A baffling instance of a safety device being overriden by the man who conceived it is shown in an accident at Paris in 1902. Augusto Severo designed and built a motor-driven navigable balloon. He had taken the double safeguard approach and incorporated two safety valves to relieve the pressure of the hydrogen as it expanded at the high altitude he expected to reach. At the last moment, however, he dispensed with one of the safeguards and sealed up a relief valve, claiming it served no purpose. With a mechanic aboard he took off in a steep and rapid climb. One of the two aeronauts seems to have panicked at the precipitate take-off and jettisoned some ballast, his action of course only increasing the speed of ascent. The single valve could not cope with the sudden expansion of the hydrogen and the balloon exploded, a spark from the engine having ignited the escaping gas. Both men were killed when the balloon crashed to the ground in flames.

MOST HEAVY MACHINERY today can boast an emergency stop button and it is a safety device essential to the protection of those using the machinery. But consider the Domino Effect of the following sequence of events, when the button itself was subject to an accident. It all happened at the ethylene plant of a large chemical works.

An operator tripped over and in putting out his arm to break the fall he knocked the emergency shutdown button on one of the two compressor steam turbines. It immediately shut down. The other compressor was consequently overloaded and it tripped out, causing the whole of the naphtha cracking plant to shut down. The work's power station was thus suddenly deprived of a big user of steam and power and was unable to reduce the firing on its two boilers quickly enough. One boiler consequently closed down automatically. The remaining boiler could not cope with the resulting load, so that it too shut down. The other plants on the site were now deprived of all steam and power and were, naturally, also out of action. In the end, the whole site was brought to a standstill. A great deal of production was lost and it took a week to bring all the plants back on line.

After such a debacle, it need hardly be said, a guard was put around all emergency stop buttons on the site.

It has been calculated that by 1850 three million horses would have been needed to produce the power then being generated by steam. Boiler explosions were commonplace and led eventually to the employment of official Boiler Inspectors, but their benevolent activities were not always taken in the right spirit. From 1868 comes a report of one Inspector's experience:-

> '.. with a very slight blow the Inspector struck a chisel through almost every plate (along the right hand seating wall). The owner being from home the inspector waited his return, and described to him the dangerous condition of the boiler. Instead of expressing satisfaction that an explosion had been averted by this timely detection of the defects, he became much excited and in language more forcible than polite expressed his displeasure at the Inspector having come there to make holes in the boiler.'

Another instance of a Boiler Inspector's mission being frustrated was the interesting history of a boiler in Glasgow. It was installed, second-hand, in a distillery in the 1860s and discarded in 1880 after an Inspector had been able to knock a hole through the bottom. It was then picked up by another Scotsman who patched the hole and worked the boiler continuously until 1914 when another Inspector knocked a hole through the plates. Only the threat of legal action prevented the thrifty owner from patching the boiler up again.

But there was nothing wrong with the boiler on Locomotive No. 157 of the London and North Western Railway when it blew up in a siding at Wolverton on 26th March 1850. Nor was there anything wrong with the pressure relief valve on the boiler — that is until human folly thwarted its purpose.

The boiler fire had been kindled in the early hours of the morning and a steam pressure of 60 lb/sq.in. reached shortly afterwards, resulting in the relief valve lifting and steam escaping. A fitter's mate, working on an adjacent engine, found the noise of this escaping steam somewhat annoying and screwed the relief valve down tightly to stop the escape of steam. At 6 a.m. the driver of No. 157 arrived and noticed that steam was escaping strongly from the relief valve but did not realise that the pressure was now 95 lb/sq.in. He oiled the engine and fifteen minutes later, when he opened the regulator, the boiler barrel exploded and opened out from the bottom seam. The boiler shell uncurled into a flat sheet as it flew through the air.

The driver was unhurt, but the fitter's mate, who was blamed for the accident because he had screwed down the relief valve, was scalded and had one of his sensitive ears blown off.

AS WE HAVE SEEN, the whole idea of a lightning conductor is to attract the electrostatic charge and bring it safely down the conductor to earth. But Murphy has a finger in every pie as is shown in the following incident, when the lightning ran *up* the conductor.

It happened in 1975 in the Netherlands. A steel tank to contain a flammable material had been constructed, with a concrete cover topped by about half a metre of earth. The tank was provided with a lightning protection system which ended in six earthing plates, and it was therefore a playful stroke of fate when a bolt of lightning avoided a nearby radio mast, struck a 20 m high willow tree, travelled down the trunk into a root that touched one of the earthing plates. The electrical impulse then passed up the earthing plate and conductor into the tank where it ignited the contents and set off an explosion which blew off the concrete covering and severely damaged the tank.

THE SECOND LORD LECONFIELD was a caring family man. Among his worries was the provision of safe drinking water when they were all in residence in London, in those years beset by typhoid and cholera. A cart daily trundled up to town laden with barrels of fresh spring water from his estates at Petworth in Sussex. This precaution continued until January 1895, when one of his nephews died of typhoid fever contracted at Petworth.

BLIGHT ALSO ATTACKED the good intentions of Mr. W. R. Wyndham, another member of the same family. Before the Great War the Petworth estate boasted its own horse-drawn fire engine which had a great brass boiler, capable of producing an enormous pressure of water when a fire was lit inside it. On the only occassion when this fire engine was used an enthusiastic Mr. Wyndham galloped with it to the scene of a small fire in one of a row of cottages, only to find the situation well under control. Undaunted, he directed so fierce a jet of water on the cottages that all of them collapsed.

In earlier centuries, before the formation of fire brigades, putting out fires was a duty imposed on all able-bodied citizens, who were summoned to the task by the ringing of church bells or other rousing alarms. In Scotland drums were frequently used to call out the amateur firemen, but this admirable system could be frustated by the hoax call. Leith, near Edinburgh, was a port and as such was wide open to the 18th century activities of the press gangs. It is related that on one occasion when the press gangs were suspected of being on the prowl the sensible residents kept a low profile and remained indoors. The gangers, however, were totally unscrupulous and beat out the fire call on drums; they promptly impressed all the conscientious young men who came rallying to the alarm.

This trick, of course, evoked the Matilda Syndrome (For every time she shouted 'Fire!' They only answered 'Little Liar!'), and after a serious fire in Edinburgh in May 1755 the *Scots Magazine* sadly noted:-

> 'The working people on hearing the fire drum suspected it to be a stratagem used by the press-gang, who had used such an one sometime before at Leith; and therefore sufficient help was not so speedily got as usual in such cases.'

AN ITEM OF SAFETY EQUIPMENT that did not quite reach the makers' expectations for it was the classic cork lifebelt. In itself it was an excellent idea and undoubtedly saved many seafarers from sinking when worn in the water. But for anybody wearing one who had to leap from a burning or sinking ship there was an added hazard. The thick but buoyant cork did not, of course, initially submerge with the same velocity as the wearer's body, and more often than not a broken jaw was added to the other miseries of being consigned to the open sea.

WE KNOW ABOUT STABLE-DOOR LEGISLATION and there is also half-cock legislation which, however well-intentioned, nevertheless leaves so many loopholes as to render it virtually useless. Our example comes from the 19th Century.

One of the hazards the mercantile nation had to deal with was the overloading of her trading ships. The confident Victorians were the ones to face the problem and there were calls for a loading line on the ship's hull to indicate the safe limit of submersion. Samuel Plimsoll, called with reason The Sailor's Friend, was a Member of Parliament and campaigned hard for load lines. He was successful, and in 1875 his lines were made compulsory. It seems, however, that the State did not wish to appear too interventionist, and naively left it to the individual ship owners to decide where to draw the line, or rather to paint it. In theory they could paint it half-way up the funnel if they chose, and unscrupulous owners continued to send dangerously overladen ships to sea for another fifteen years. They were rumbled in 1890 and a further Act was passed to make the proper use of Plimsoll lines enforceable.

**HOWEVER, AS MIGHT BE EXPECTED** in this capricious life, it is when we embark on a carefully planned safety crusade that we are most likely to go adrift. Not that such a crusade should be abandoned — far from it. But it is as well to bear in mind that even in the area of safety programmes strange and rare stumbling blocks are to be encountered. Take the following, for example. A management meeting decided on a scheme to persuade its work force to wear safety helmets; a five pound note would be given to the first two employees seen, in randomly selected groups, wearing their hard hats. But they had reckoned without the archaic regulations of the bureaucrats, and the firm's accountant had to report back that the Inland Revenue would lean on the winners for its share of the £5.

**SAFETY GOGGLES** are another item of protective clothing the wise employer will provide. The wise employee will wear them. But sadly there is a dearth of wisdom in some workplaces and many employers have had to introduce programmes specifically designed to deal with the problem. A chemical company laid on a demonstration for its laboratory staff to underline the dangers of not wearing goggles. A butcher generously supplied a sheep's eye which was placed on a plate to stare up at the assembled staff while the safety officer gave his sales pitch. Then came his *coup de théatre*, as he whipped out a bottle of one of the chemicals made at the plant and poured a small amount on to the lonely sheep's eye. It instantly turned white all over. More than half the spectators, equally white, raced from the room and clocked off sick.

A similar own goal was scored by an engineering company in 1976 when a film was shown to bring home to the workforce the blood-chilling effects of not wearing safety goggles. So realistic was the film that a dozen fainting members of the audience had to be helped from the room; one particularly squeamish welder was so affected that he gashed himself falling off his chair in a faint and needed seven stitches to restore him to full working order. The safety officer later announced that the film was being withdrawn on the grounds that it constituted a safety hazard.

AT A CHEMICAL PLANT recently a safety incentive scheme was devised. Six video recorders were to be put into a year-end draw, with one recorder being taken out of the draw for each lost time accident during that year. Sadly there were exactly six such accidents by the end of December and no recorders were left to be drawn for. Undaunted, management tried another tack, and with six unused video recorders on their hands they announced a new scheme. At the end of each month when there had not been a lost time accident a recorder would be drawn for. The names of all employees at the site were to be included and when the lucky winner of the first draw, made by the trades union convener to ensure fairness, turned out to be the Works Manager himself the whole idea of a draw was scrapped.

COOKING THE SAFETY BOOKS is a singularly unattractive piece of chicanery. The practice can be relatively harmless, albeit dishonest, as when an operator fell off his bicycle just as he had entered a plastics plant. In order not to have the accident attributed to the site — possibly because of some safety competition — the gateman pulled the bicycle outside the gate and claimed the accident had happened off the company's premises.

More serious though was an episode at a plant where lead and chrome salts were produced for pigments. An enthusiastic new medical officer expressed concern at the large number of operators showing signs of lead poisoning and discussed with the Works Manager ways of reducing the exposure of the men to the toxic chemicals. To the great delight of the doctor a dramatic improvement in the statistics followed. It later emerged, however, that the Works Manager had introduced a procedure of his own; he simply sacked any operator showing early signs of lead poisoning.

FROM THE UNITED STATES OF AMERICA recently came a disheartening press report. Forty-five people were injured — one of them an elderly woman whose back was broken — when a safety drill was enacted in an unsuccessful attempt to prove to the Federal Aviation Authority that a new jumbo jet could be evacuated in 90 seconds. The simulated emergency was all too realistic for the 410 volunteer pseudo-passengers. One of them, whose foot was broken, was quoted: 'Everyone was screaming, pushing and yelling... there were so many people behind you that you got shoved on to the chute.' Despite this unfortunate first test, 'a second drill went ahead and there were several more injuries.'

A recent safety exercise at a plant in the north of England also used outsiders as casualties. It was to have been a carefully planned, unheralded operation involving the emergency services, but a disenchanted participant wrote afterwards:-

> 'We were simulating a leak at a gas plant, so it was supposed to be a disaster. Unfortunately, it turned out to be a bigger disaster than we had bargained for.
>
> 'It didn't start well. The actors pretending to be casualty victims arrived three hours late and there was some sort of argument about catering and who would get 'star billing'. Then a police car drove up to the south gatehouse and the officer inadvertently alerted the staff that a secret exercise was about to take place. It hardly mattered anyway because, when the chemical emission alarm was sounded, neither the police nor the fire brigade heard it, and half an hour later someone had to phone them up to tell them it had gone off.
>
> 'Things got worse. A fireman fell through an open staircase, a volunteer had a severe asthma attack on a high gantry, and three other volunteers got hypothermia after being drenched with fire hoses, and we had no option but to abort the exercise. God alone knows what will happen if there ever really *is* a leak at the plant. It's the last time I work with actors.'

WRITTEN INSTRUCTIONS are a fertile ground for misunderstanding. A drug trade journal in 1988 described in great detail the 'Management of Serious Paracetamol Poisoning'. In these instructions it was laid down that 'the patient should be nursed at 30°–40°, given 100ml...' Two months later the journal issued a correction: 'Under the section *Encephalopathy* we said the patient should be nursed at 30°–40°. This referred to the angle in bed — not the temperature.' How disappointing that we have not been told what happened to patients nursed at a 40° temperature.

And a warning sign put up recently by well-meaning British Rail staff at Blackfriars Station would have perplexed those who take words at their face value. It read:-

CAUTION — UNEVEN FLAW

A tragic coincidence thwarted the good intentions of a company when it embarked on a new programme in pursuit of excellence with the aim of reducing workplace accidents. The slogan fixed on was *Reaching New Heights*. A booklet was prepared and on its cover, under the slogan, was a drawing of an American space shuttle soaring above Europe but with a strange star burst behind it. The programme was launched on New Year's Day, 1986. A fortnight later the space shuttle Columbus exploded on lift off.

Equally unfortunate was a brochure produced to extol the virtues of a nautical sprinkler system. The marine oil tanker featured in the illustration later caught fire and sank. Indeed, the use of illustrations is often a trap for the thoughtless. The Health and Safety Commission have as their very appropriate symbol a pair of cupped, caring hands. Someone, somewhere, had the idea of putting this symbol on the front of some tee-shirts and getting a group of the female staff to model them for a photograph in the house magazine. Feminists to a man, they all refused.

# CHAPTER 13

# ODDITIES

HERE IS A RAG-BAG of curiosities to suit all tastes, more or less connected with accidents and matters of safety. They span the centuries, and those who treasure the sometimes bizarre thinking of our forefathers will appreciate the first offering, from as recent a date as 1897.

IN THAT YEAR in the United States Edwin J. Goodwin had been studying his Old Testament. In II Chronicles 4, 2, he read:-

*Also he [Solomon] made a molten sea of ten cubits from brim to brim, round in compass, and five cubits the height thereof; and a line of thirty cubits did compass it around about.*

From this he made a simple calculation and proposed in the Indiana House of Representatives 'A Bill Introducing a New Mathematical Truth' to restore the value of pi ($\pi$) to its Biblical value of 3. Oddly enough, it passed the First Reading, and only the fortuitous presence of mathematician C. A. Waldo prevented the Bill passing into law.

AT ABOUT THIS TIME, dentists were experimenting with new materials for false teeth. Celluloid teeth were tried for a while but, though cheap, were found to be highly flammable. It is reported that at least one wearer was severely burned when his teeth caught fire while he was smoking a cigarette. They should have stuck to 'Waterloo teeth', so called because dentists in those days, anticipating Donor Cards, were in the habit of recycling human teeth collected from battlefields.

THE HISTORIAN HERODOTUS (5th Century BC) says that King Amasis of Egypt sent a gift of 1000 talents of alum for the rebuilding of the temple at Delphi which had been destroyed by fire in 548 BC. Alum is a useful fireproofing material, so this was a good-natured gesture. Victorian ladies mixed alum with starch to fireproof dress fabrics; a necessary precaution when all too many swaying crinolines caught fire as their wearers floated past the open fireplaces.

FIREPROOFING WAS ALL THE RAGE in those days. In 1850 you could buy fireproof ink (made from platinum chlorate, oil of lavender, varnish and lamp black) and use it to write on fireproof paper (wood pulp and size treated with borax and asbestos fibre).

The origin of the phrase 'setting the Thames on fire' is believed to refer not to the river but to a piece of equipment used in flour mills called a *temse*, a kind of sieve. When flour was sifted through this *temse* by hand, a man could work so hard at it that the friction would cause the wood to smoulder and ignite. A slow worker could thus not set the *temse* on fire.

SLOW AND INEFFICIENT WORKERS were not unknown in ancient Rome, where slaves (who else) were employed as fire fighters. They were called the *Familia Publica* and a tribunal of magistrates, the *Tresviri Nocturni*, was responsible for them. It is in the records that in 300 BC complaints were made that firemen were slow in coming to a fire. They were indeed a pretty inefficient bunch, and in the year 6 AD, after a disastrous fire, Emperor Augustus instituted the Corps of Vigiles, a highly professional body of firemen, who even had a right of entry into any building to check on the safe use of fire for cooking, heating or lighting. They disappeared from history with the fall of the Roman Empire.

But other fire brigades came into being. At Burton in Derbyshire around the turn of the century twelve young women constituted the Ladies Fire Brigade, equipped with a manual pump and lowering lines. Although they wore men's tunics over long skirts we doubt that they copied some of their male counterparts who, when pumping water, rhythmically chanted 'Beer oh! Beer oh!' to let the gaping crowds know what was expected from them.

An alternative to the manual pump was a fire extinguisher, called the Phillips Patent Fire Annihilator, which won a prize at the Great Exhibition of 1851. In 1852 the London factory where it was manufactured burned down and went out of business.

APPORTIONING BLAME for accidents and the imposition of appropriate punishment have, of course, varied over the centuries. In his *Penitentials* Archbishop Theodore tells us:-

> 'By Saxon law, if the man put the heavy cauldron of water on to the fire and it boiled over and scalded the baby lying by the fire, the woman was held to blame for leaving her helplessly swaddled baby in such a dangerous position.'

Turkey cannot have been a fun place to live in during the 18th century for those who caused fires, deliberately or not. It was reported in January 1783:-

> 'A fire broke out near the mosque of the Sultan Achmet, at Constantinople in December last, which reduced ten palaces to ashes. The new Grand Vizier ordered the throats of ten or twelve persons to be cut, on being accused of having concealed combustible materials in different quarters of the city.'

WORDS CAN BE DEVIOUS little devils in the wrong hands. A failure at a chemical plant in the USA caused a brown cloud of nitrogen peroxide to pass over the neighbourhood, generating much adverse comment. The public relations people were wheeled in to allay public anxiety by describing the gas as 'rusty steam'. We wonder how the PR men would have described the spectacular purple cloud of iodine vapour which hovered over the City of Hull in November 1989? And we can instance factory inspectors' reports, which are not of course intended to be models of literary style. One noted that an accident happened 'because the cock was stiff and it could not be operated,' while another report on a steam engine proposed that the 'governor's balls should be guarded.'

BLAME FOR A DEATH was attached to a statue in the 10th century, when the County of Cheshire was suffering from a disastrous drought. Lady Trawst, wife of the Governor of Hawarden, prayed for rain before a statue of the Blessed Virgin Mary whereupon a tremendous thunderstorm broke out. The vibrations from the immense rolls of thunder loosened the statue from its niche and it and fell on Lady Trawst, crushing her to death. The statue was tried for murder, found guilty and sentenced to be hanged. But it was found impossible to hang the heavy statue and the court looked for another suitable punishment. The statue was eventually tied to a cross and left on the banks of the River Dee to drown.

The statue must originally have been in a church, for otherwise it would surely have been forfeit as a deodand (from *Deo dandum*, a gift to God). The principle in English law, dating from the Middle Ages, was that any chattel that caused a death had to be offered to God in expiation; in practice it usually went to the Crown to be used for pious purposes such as alms giving. Early Coroners Rolls record many instances where, for example, a horse that threw its rider to his death was seized by the Sheriff as a deodand.

IN 1765 THE GREAT JURIST Sir William Blackstone in his *Commentaries* stated:-

'If a man falls from a boat or ship in fresh water, and is drowned, it hath been said that the vessel and cargo are in strictness of law a deodand.'

This view was evidently held as late as 1841 by a Coroner's Inquest enquiring into the deaths of eight victims of a railway accident on the Paddington to Bristol line. The engine *Hecla*, with two third-class carriages, a parcel van and seventeen goods wagons ploughed into a great mound of earth dislodged from a cutting by abnormally heavy rains. The weight of the goods wagons crushed the two passenger coaches. The jury at the inquest declared a deodand of £1,000 on *Hecla*, payable to the Lord of the Manor where the accident had happened (much more sensible than giving him the actual railway engine to use for pious purposes). However, the Board of Trade exonerated the railway company, who successfully appealed against the deodand. This was probably the last time the ancient Common Law was invoked, and it was formally abolished in 1846.

IN MEDIAEVAL TIMES — and even later — animals were put on trial for causing death and other mishaps, and were executed if found guilty. A pig was hanged in Normandy in 1394 for eating a child, and in Germany in 1499 the trial of a bear accused of molesting villagers was delayed for legal argument. But the most mind-boggling of such criminal proceedings was a trial in 1474 at Basle in Switzerland. A cockerel was found guilty of laying an egg 'in defiance of the natural law.'

STILL IN THE ANIMAL WORLD, a shocking accident happened to a red setter a few years ago when he cocked his leg at an illuminated No Through Road sign. An electric current passed up the stream, shot through his tender parts and up the metal dog chain into his owner's arm. The road sign disintegrated and the stitches of the unfortunate dog's hip operation melted.

ELECTRICITY WAS ALSO at the heart of a mystifying experiment reported in the *Gentleman's Magazine* in 1747. One Thomas Yeoman, FRS, an engineer, was demonstrating some of the effects of the newly discovered energy source:-

'Mr. Yeoman having electrify'd a man, while a hole was open in his vein, the blood flow'd then much faster, and slower on ceasing to electrify.'

And a few years later, on 17 February 1753, John Wesley noted in his Journal:

'From Dr. Franklin's Letters I learned... that the electrical fire, discharged on a rat or a fowl will kill it instantly; but discharged on one dipped in water, will slide off, and do it no hurt at all. In like manner, the lightning which will kill a man in a moment, will not hurt him if he be thoroughly wet. What an amazing scene is here opened for after ages to improve upon!'

Amazing indeed.

AROUND THE TURN OF THE CENTURY the Commandant of the British Army's Balloon Factory at Farnborough designed a pressure airship whose envelope was constructed of five layers of goldbeater's skin, a material made from the intestines of oxen. Alas, five layers of this exotic fabric proved too heavy and the airship failed to rise from the ground.

THE WATCHDOGS IN BRUSSELS who worry about the safety of our food and drink may well muse over the next two items. The first was reported in *The New Annual Register* on 26th April 1785:-

'The grand jury of Lancashire, at the last sessions, presented Samuel Higginson, of Manchester, for having on the 4th of November preceding, breathed into, blown and inflated the loin, heart, liver, and lights of a calf then newly slaughtered, to the intent of giving them the appearance of large, fine and wholesome victuals, by which means the said loin, etc. became corrupt, nasty, fetid, and unwholesome, and unfit and unsafe to be eaten by his Majesty's subjects; and for having afterwards hung up and exposed the same for sale, contrary to the statute in that case made and provided.'

The statute referred to was probably that enacted in 1266 prohibiting the sale of 'corrupted wine' or any meat, fish, bread and water 'not wholesome for man's body' or 'that was kept so long that it loseth its natural wholesomeness.' This law continued in force until repealed in 1844.

The second item concerns a particular Spanish wine from Cadiz which was found to make a very good brandy after the distillation process. When the vat of wine was emptied, the skeleton of a man was found in the bottom. He was eventually identified as a worker who had disappeared some time previously and for whom there had been an extensive but unsuccessful search. It was believed that he had fallen into the vat at the fermentation stage. It is related that subsequent vats of wine have part of a cow added (we do not know which part) to give the wine the required condition.

JUDGING BY the numerous references in his diary to nourishment, Samuel Pepys must have been a great gourmet. But he had wide-ranging interests and wrote on the 26th June 1663 about a dinner he had attended with some naval cronies:-

'At table, discoursing of thunder and lightning, they told many stories of their own knowledge; of their masts being shivered from top to bottom, and sometimes only within and the outside whole. But among the rest, Sir Wm. Rider did tell a story of his own knowledge, that a Genoese Galley in Legorne road was struck by

thunder so as the mast was broke a-pieces and the shackle upon one of the slaves was melted clear off of his leg. Sir Wm. went on board the vessel and would have contributed towards the release of the slave whom Heaven had thus set free, but he could not compass it and so he was brought to his fetters again.'

A YEAR BEFORE, the 17th century Quaker Henry Smith described a mildly apocalyptic vision he had:-

'As for the great city a fire was kindled therein and there was none could quench it. And the burning thereof was exceeding great. All the tall buildings fell, and it consumed all the lofty things therein and the fire searched out all the hidden places.'

That was written in 1662, four years before the Great Fire of London.

Henry Smith was not the only one to foresee the collapse of man-made structures. The famous engineer Marc Isambard Brunel (1769–1849) predicted that the first suspension bridge over the River Seine would fall down. He made his judgement by merely studying the drawings, and the bridge did indeed collapse. Like father, like son. The even more renowned engineer Isambard Kingdom Brunel made a similar prediction when at school in Hove on the South Coast of England in the 1820s. He had carefully watched the construction of a building opposite his school and pronounced that it would collapse. The building fell down that night and young Isambard cleaned up on the bets he had taken.

A MISHAP WITH A SUPERNATURAL FLAVOUR is reported from the United States. A retired nursing assistant was hit on the head by a falling plank of wood at a department store and claimed damages because the bump on her head had ended her psychic powers. She had first discovered these powers while recovering from a wrist injury. Perhaps before she retired she had nursed the American human cannonball, Elvin Bale, who is said to have 'landed in hospital' after he was shot from a circus cannon and missed his soft landing cushion area.

READERS OF CHARLES DICKENS will appreciate the horror of human spontaneous combustion as described in *Bleak House*. In his own preface to the novel, Dickens, anticipating scepticism, refers to a famous instance which happened at Rheims early in the 18th Century and which was investigated by one of the foremost French surgeons of the time. 'The subject was a woman, whose husband was ignorantly convicted of having murdered her; but, on solemn appeal to a higher court, he was acquitted, because it was shown upon the evidence that she had died the death to which this name of Spontaneous Combustion is given.'

*Stop me not, but onward let me jog,*
*For I am the London fireman's dog.*

This was inscribed on the collar of Chance, the mongrel mascot of the London Fire Engine Establishment. He was well known to the crowds of bystanders at fires in the 1830s and usually got as loud a cheer as the firemen themselves when he raced onto the scene with the thundering horse-drawn fire engines. Charles Dickens knew of him. Writing as Boz, he mentioned the burning down of the Palace of Westminster in October 1834 and noted that one of the MPs 'and the celebrated fireman's dog were observed to be remarkably active at the conflagration of the two Houses of Parliament — they both ran up and down, and in and out, getting under people's feet, and into everybody's way, fully impressed with the belief that they were doing a great deal of good, and barking tremendously.'

ANOTHER IN THE WORLD of firefighting to have his own little piece of verse was Eyre Massey Shaw, redoubtable head of the London Metropolitan Fire Brigade at the end of the 19th century. He was a friend of W. S. Gilbert, who wrote him into one of the songs in *Iolanthe*:-

*Oh, Captain Shaw!*
*Type of true love kept under!*

*Could thy brigade*
*With cold cascade*
*Quench my great love, I wonder!*

It has been alleged that Gilbert based this tribute to his friend on an old music hall song praising an earlier London fire chief, James Braidwood.

*My heart's on fire!*
*Not all the Fire Brigade could*
*Subdue the flames*
*Though led by Mr. Braidwood.*

Braidwood was in fact leading his fire brigade at a disastrous fire in London in 1861 when a burning warehouse wall collapsed on him. His body was not recovered until three days later.

THE TOUGHEST OF ALL unlucky men must surely be the American politician David Kennison. It is related of him that his hand was blown off in the battle of Sackett's Harbour in the war of 1812, that his skull was fractured by the falling branch of a tree and that both his legs were severely splintered by a misfired cannon while he was a member of the Massachusetts State Militia. The fractures healed but for the rest of his life he suffered from festering sores on his legs. Later on he was murderously kicked in the face by a horse and permanently disfigured. He battled on and died in 1851 at the great age of 115, the oldest surviving person to have taken part in the Boston Tea Party.

ANOTHER SURVIVOR is commemorated by an inscription, now almost obliterated, in the churchyard of Keyshoe, in Bedfordshire. It reads:-

'In memory of the mighty hand of the Great God and our Saviour Jesus Christ, who preserved the life of William Dickens, April 17th, 1718, when he was pointing the steeple and fell from the ridge of the middle window in the spire, over the south-west pinnacle. He dropped upon the battlement, and there broke his leg and foot, and drove down two long coping stones, and so fell to the ground with his neck upon one standard of his chair, when the other end took the ground. He was heard by his brother to say, when near the ground, 'Christ have mercy upon me! Lord Jesus Christ, help me!' It is added that he died November 29th, 1759 aged 73 years. The height from whence this person fell was not less than 132 ft and his leg and foot were exceedingly fractured but his injury in other respects was so trifling that he not only lived more than 40 years afterwards but within seven months from the time of his fall, he was capable of ascending the steeple a second time and he finished the pointing of the spire.'

CONCERN FOR PRODUCTIVITY was alive and well in the 15th century. About 1420 the engineer Filippo Brunelleschi was charged with completing the famous dome of the cathedral at Florence. It is related that at one point he realised his workmen were losing too much time climbing up and down the ladders of the scaffolding at meal times and he therefore set up a canteen on the scaffold — not, it must be observed, a particularly safe place to eat spaghetti. Charles IX of Sweden used less benign methods to increase the output of his miners. 'Beat their heads in,' he wrote to his bailiff in 1608, 'stick a spear in their sides, if you should kill one it does not greatly matter.'

INSURANCE COMPANIES with heavy burglary claims might care to liaise with the building industry over a safety measure taken at Lady Stair's House, Royal Mile, Edinburgh, built in 1622. Possibly the lady's name suggested the idea. In this house the risers of the stairs were not all built to the same height, so that persons unfamiliar with the house might betray their presence by stumbling. [N.B. The author has personally measured the stairs and can reveal to feloniously minded visitors that the eleventh stair is 8½ inches high whereas the others are only 6¾ inches.]

Still in Scotland, it is perhaps fitting that the oldest vehicle surviving there should be a hearse. It is preserved in the National Museum of Scotland and is called the Bolton Hearse. Parts of it date back to the mid-seventeenth century.

A couple of centuries ago it was believed that the shock of a sudden fright would cure whooping cough and other childhood ailments. It didn't always work that way, and in August 1736 a criminal trial was held at Rochester in Kent

'... of a Soldier who pretended to cure a Boy of an Ague, and thinking to frighten it away, by firing his Piece over the Boy's Head, levell'd it too low, and shot his Brains out...'

Another curious medical belief was that held by the naturalist and physician Martin Lister. In his *Dissertation on the Pox*, published in 1694, he advanced the remarkable theory that syphillis was first transmitted to man, in tropical America, by the giant iguana lizard.

**Denys Papin**, who died in 1710, had an idea for harnessing steam energy. He invented what contemporaries called his 'bone digester' — so named because he first demonstrated its power at a dinner party at the Royal Society, where the food was cooked in the new apparatus. Diarist John Evelyn attended Papin's dinner and wrote on the 12th April 1682:-

> '... the hardest bones of beef itself, and mutton, were made as soft as cheese... This philosophical supper caused much mirth amongst us and exceedingly pleased all the company.'

The 'digester' on that occasion served as a pressure cooker, but it became known to industry as the autoclave. It was constructed of bronze and fitted with a weighted lever and valve. This was the first reference to a safety valve. Papin later designed what is said to have been the first technically sound steam engine, but oddly enough he did not incorporate his safety valve in the specification.

**John Evelyn**, incidentally, was a peaceable man. He confided in his diary on 16th March 1687:-

> 'I saw a trial of those devilish, murdering, mischief-doing engines called bombs, shot out of the mortar-piece on Blackheath. The distance that they are cast, the destruction they make where they fall is prodigious.'

During the First World War some Austrian troops contributed their share of prodigious destruction. They were stationed on the Alps and in December 1916 fired their cannons in a training exercise, setting off an avalanche which is said to have killed 2,000 people.

**In Victorian times** some street lighting was provided by Webb's Patent Sewer-Gas Destructor, which made use of what was referred to as 'the gases of putrefaction' piped up from the sewers of London. Should we perhaps revert to this source of methane gas as an infinitely renewable energy source?

Gas powered too were the first traffic lights in London. They were installed in 1867 in New Palace Yard in Westminister and were not a roaring success. Two years later they exploded, seriously injuring the policeman who was responsible for operating them.

**On the subject of explosions**, we can reflect upon yet another hazard of smoking. In older marine oil tankers the accommodation section was set amidships, and consequently the waste pipe from the lavatory passed down a concentric pipe through the central cargo tank. It was not unknown for these pipes to corrode. Oil or vapours would then enter the waste pipe but, being heavier than air, would stay at the bottom of the pipe. One misguided officer, having had a surreptitious smoke, dropped the cigarette end down the lavatory and immediately sat down. There was only a small explosion but it clearly affected the officer; ever since this episode he has had a pronounced stutter.

In June 1989 nineteen people were drowned when their sailing boat capsized off Zanzibar. She was called *God Save Us*.

The Almighty's assistance was also invoked at Sherborne Abbey in the Middle Ages when a bell was hung, to be tolled as a warning of fire, bearing the following inscription:-

*'Lord, quench this furious fire.*
*Arise, run, help, put out the same.'*

IS IT NOT IRONIC that, long after Prometheus stole fire from Olympus and was punished by the gods for his theft, Olympia in London should have been chosen in 1975 as the venue for an international convention of Fire Chiefs, together with an exhibition of the very latest in fire fighting equipment. The gods must have been angered by this fresh instance of presumption, for the hotel where all the Fire Chiefs were staying then caught fire.

The gods also punished the Royal Society for the Prevention of Accidents — for what act we do not know — by collapsing their grand display stand at an exhibition in Harrogate in 1968.

LOPPING OFF THE RIGHT EAR was the punishment for certain Mediaeval misdemeanours and a criminal past was thus plain for all to see. It is therefore not surprising that one young man was obliged to publish the following:-

'20 May 1303, Roxburgh. Notification, lest sinister suspicion arise hereafter, that the right ear of William son of John le Noble of Laghton was torn off in his minority by the bite of a pig.'

ABOUT 1550 the learned Georg Bauer, better known by his Latinized name of Georgius Agricola, wrote *De Re Metallica*, a scholarly work which detailed his considerable experience of and observations on the state of mining and metallurgy at that time. He also wrote a perfectly serious treatise on the problem of gnomes and how to eliminate them from mines.

But the Little People lived on in mines for at least another century. The *Philosophical Transactions of the Royal Society*, 1700–1701, included a paper by a Captain Sammuel Sturmy, in which he reported:-

> 'Upon the 2nd of July 1699, I descended by ropes affix'd at the top of an old Lead-Ore Pit, 4 Fathoms almost perpindicular; and from thence 3 Fathoms more obliquely, between two great Rocks, where I found the Mouth of this spacious Place; from which a Mine-Man and my self lower'd ourselves... and the Mine-Man went up the Ladder to that Place, and walk'd into it about 70 Paces, till he just lost sight of me, and from thence chearfully called unto me, and told me he had found what he had look'd for, a Rich Mine. But his Joy was presently turned into Amazement, and he returned affrighted by the sight of an Evil Spirit, which we cannot persuade him but he saw, and for that Reason will go thither no more.'

THE SIZE OF SUNDAY NEWSPAPERS has grown considerably and it is not surprising that the weight of these papers should give rise to some concern. In the United States of America the weight is greater than over here, and because they are tossed into house doorways instead of being rammed into letterboxes the hazard is compounded. It is doubtful if this aspect of the delivery practice was appreciated by the newspaper proprietors until 1985. In that year a women successfully sued the *Los Angeles Times* for the loss of her dog, killed when a newspaper hurled by the delivery boy landed on the unsuspecting creature.

SINCE THE NINTH CENTURY St. Barbara has been revered as the protector of miners, gunners and military architects. Legends dating back to the ninth century place her in many places — Egypt, Rome, Tuscany. According to the Golden Legend she was a young girl of considerable beauty, who was locked up in a tower by her father, Dioscurus, to discourage attentive young men. During this captivity she was converted to Christianity. Her pagan father denounced her to the prosecuting authorities, and when she refused to renounce her faith Dioscurus was ordered to put her to death. This he did and was promptly struck dead by lightning. Because of her father's fate, she has been invoked against lightning, which danger was later extended to include explosions and anything that might cause one.

Her cult was widespread. On Spanish naval ships the magazine where ammunition was stored was known as the *Santa Barbara*, and in tunnelling operations in the Alps when explosives were used it was customary to place a statue of Barbara in a niche cut into the rock at the tunnel opening.

St. Barbara was also the patron of the Royal Regiment of Artillery. She would have been keeping a watchful eye on events in 16th Century India, where a supergun had been set in place at Tanjore. The barrel was 25 ft long and 3 ft in diameter; the whole gun weighed nearly 40 tons. The workmanship is considered to have been superior to contemporary European techniques. There is a legend that it was only fired once, for such was its awesome size that no gunner could be found to venture near it and a train of gunpowder, two miles long, had to be laid to the touch-hole. When the gun was eventually fired the force of the shock was said to have seriously damaged the health of all those living near by.

Another remarkable gun, now in the Musée de la Marine in Paris, is *La Consulière*, named after the unhappy French Consul in Algiers at the time of the Pirate Wars of 1830. He was fired from it at the French fleet in an act of derision by the pirates who had captured Algiers.

Somewhat earlier, King James II of Scotland was killed by the accidental bursting of one of his own cannon during the siege of Roxburgh Castle in 1460. He had probably ignored a common precaution taken by gunners of those days — to stand ten paces to the rear and one to the left of the gun on firing, this being considered the safest place from which to escape flying pieces of metal.

It is sad that Barbara has now been expelled from the Calendar of Saints, but there is another saint of possible interest to safety buffs. At a church in Cremona, Italy, is a statue of a man with an axe buried in his head. It is said to represent Pietro Esorcista and has been identified with St. Peter of Verona, Martyr. This St. Peter was born in 1205, joined the Dominicans, and gained a great reputation as a preacher and miracle worker. While on a journey from Como to Milan in 1252 he was attacked by two villains, one of whom clove his head with an axe. This friar from Verona should perhaps be appointed patron saint of safety officers, whose efforts to persuade workers to wear safety helmets are as fruitless as they are unceasing.

# CHAPTER 14

# IN SONG AND VERSE

THERE IS HARDLY a human experience that has not at one time or another been the subject of poetry. Accidents are no exception, and one of Milton's greatest poems, *Lycidas*, laments the death of his friend Edward King, who was drowned in 1637 while crossing from Chester to Ireland. The verses offered below reach no such heights, but they may entertain or instruct. Some have been set to music.

One of Hilaire Belloc's *Cautionary Verses* is about a boy called

GEORGE

*Who played with a Dangerous Toy, and suffered*
*a Catastrophe of considerable Dimensions.*

*When George's grandmama was told*
*That George had been as good as gold,*
*She promised in the Afternoon*
*To buy him an immense balloon.*
*And so she did; but when it came,*
*It got into the candle flame,*
*And being of a dangerous sort*
*Exploded with a loud report!*

*The Lights went out! The windows broke!*
*The Room was filled with reeking smoke.*
*And in the darkness shrieks and yells*
*Were mingled with Electric Bells,*
*And falling masonry and groans,*
*And crunching, as of broken bones,*
*And dreadful shrieks, when, worst of all,*
*The house itself began to fall!*
*It tottered, shuddering to and fro,*
*Then crashed into the street below —*
*Which happened to be Savile Row.*

*When help arrived, among the Dead*
*Were Cousin Mary, Little Fred,*
*The Footmen (both of them), The Groom,*
*The man that cleaned the billiard room,*
*The Chaplain, and the Still-Room Maid,*
*And I am dreadfully afraid*
*That Monsieur Champignon, the Chef—*
*Will now be permanently deaf—*
*And both his Aides are much the same;*
*While George, who was in part to blame,*
*Received you will regret to hear,*
*A nasty lump behind the ear.*

*Moral*
*The moral is that little boys*
*Should not be given dangerous toys.*

THE FELLING COLLIERY DISASTER in 1812 cost the lives of 92 miners. A sombre song about the accident was written by Dr. Alan Bush:-

*MEN OF FELLING*

*Tyne to Tyne to the river shore, under the bent brow of the hill, the slope of rock and scree.*
*The hill drops to the river shore, to Tyne, great roadway to the sea.*
*Here stood the Roman once, and here his alien standard showed*
*Here ruled but never conquered quite the dark fell with its crown of cloud*
*Beneath the granite slopes a treasure lay, centuries old but scarcely touched by men*
*A store of riches still untold. For masters, hope of gain, for men, the shift from dark till dawn*
*A source of want, a source of wealth, the harvest of the mine*
*Here men of Felling man the seam but deep enough they delved for it*
*Seventy fathoms down they worked the fiery pit*
*Rich, rich was the seam they wrought until with flame circled about*
*With voice of thunder through the mine. Loud, loud the angry earth cried out*
*Disaster broke the bond of cold and silent unconcern*
*No more the cost of life the wheels of factory and pit could turn*
*Through days of doubt, of hope, of joy; through years of struggle, war or*

*peace the men of Felling lived and saw the changing centuries*
*Now are their voices joined. The hills and valleys echo loud*
*A better future now they sing in friendship and in brotherhood mastered is nature's might.*
*The worst of tyranny is done and rising from the ash of time begins a greater age for Man.*

THE ARGUMENTS between those who would prevent and those who would protect rattle along at a dashing pace in this one, by an unknown author:-

*THE AMBULANCE IN THE VALLEY*

*T'was a dangerous cliff, as they freely confessed,*
*Though to walk near its crest was so pleasant;*
*But over its terrible edge there had slipped*
*A duke and fully many a peasant.*

*The people said something would have to be done*
*But their projects did not at all tally.*
*Some said 'put a fence round the edge of the cliff',*
*Some 'an ambulance down in the valley',*

*The lament of the crowd was profound and was loud*
*As their hearts overflowed with their pity,*
*But the cry for the ambulance carried the day*
*As it spread through the neighbouring city.*

*A collection was made to accumulate aid*
*And dwellers in high-rise and alley*
*Gave pounds or gave pence, not to furnish a fence,*
*But an ambulance down in the valley.*

*'For the cliff is quite safe, if you're careful' they said,*
*'And if people should slip and are dropping —*
*It isn't the slipping that hurts them so much*
*As the shock down below when they're stopping.'*

*So for years we have heard as these mishaps occurred,*
*Quick forth would the rescuers sally.*
*To pick up the victims who fell from the cliff*
*With the ambulance down in the valley.*

*Said one as his plea 'It's amazing to me*
*That you'd give so much greater attention*
*To repairing results than to curing the cause,*
*You had much better aim at prevention.'*

*'For the mischief of course, should be stopped at its source*
*Come, neighbours and friends, let us rally.*
*It is far better sense to rely on a fence*
*Than an ambulance down in the valley.'*

*'He's daft in the head' the majority said,*
*'He would end all our earnest endeavour.*
*He's a man who would shirk this responsible work*
*But we will support it for ever.'*

*'Aren't we picking up all, just as fast as they fall,*
*And giving them care liberally?*
*A superfluous fence is of no consequence*
*If the ambulance works in the valley.'*

THERE WERE SOME startling consequences of the Tay Bridge disaster. One was the reported washing up on the Norwegian coast of an Edinburgh and Glasgow third-class railway coach about six weeks after the tragedy. Another was the lament that Scotland's unofficial poet laureate felt obliged to compose. William McGonagall wrote:-

*THE TAY BRIDGE DISASTER.*

*Beautiful Railway Bridge of the Silv'ry Tay!*
*Alas! I am very sorry to say*
*That ninety lives have been taken away*
*On the last Sabbath day of 1879*
*Which will be remember'd for a very long time.*

*'Twas about seven o'clock at night,*
*And the wind it blew with all its might,*
*And the rain came pouring down,*
*And the dark clouds seem'd to frown,*
*And the Demon of the Air seem'd to say —*
*'I'll blow down the Bridge of Tay.'*

*When the train left Edinburgh*
*The passengers' hearts were light and felt no sorrow,*
*But Boreas blew a terrific gale,*
*Which made their hearts for to quail,*
*And many of the passengers with fear did say —*
*'I hope God will send us safe across the Bridge of Tay.'*

*But when the train came near to Wormit Bay,*
*Boreas he did loud and angry bray,*
*And shook the central girders of the Bridge of Tay*
*On the last Sabbath day of 1879,*
*Which will be remembered for a very long time.*

*So the train sped on with all its might,*
*And Bonnie Dundee soon hove in sight,*
*And the passenger's hearts felt light,*
*Thinking they would enjoy themselves on the New Year,*
*With their friends at home they lov'd most dear,*
*And wish them all a happy New Year.*

*So the train mov'd slowly along the Bridge of Tay,*
*Until it was about midway,*
*Then the central girders with a crash gave way,*
*And down went the train and passengers into the Tay!*
*The Storm Fiend did loudly bray,*
*Because ninety lives had been taken away,*
*On the last Sabbath day of 1879,*
*Which will be remember'd for a very long time.*

*As soon as the catastrophe came to be known*
*The alarm from mouth to mouth was blown,*
*And the cry rang out all o'er the town,*
*Good Heavens! the Tay Bridge is blown down,*
*And a passenger train from Edinburgh,*
*Which fill'd all the people's hearts with sorrow,*
*And made them for to turn pale,*
*Because none of the passengers were sav'd to tell the tale*
*How the disaster happen'd on the last Sabbath day of 1879*
*Which will be remember'd for a very long time.*

*It must have been an awful sight,*
*To witness in the dusky moonlight,*
*While the Storm Fiend did laugh, and angry did bray,*
*Along the Railway Bridge of the Silv'ry Tay.*
*Oh! ill-fated Bridge of the Silv'ry Tay,*
*I must now conclude my lay*
*By telling the world fearlessly without the least dismay,*
*That your central girders would not have given way,*
*At least many sensible men do say,*
*Had they been supported on each side with buttresses,*
*At least many sensible men confesses,*
*For the stronger we our houses do build,*
*The less chance we have of being killed.*

THE FOLLOWING BALLAD about the accidental shooting of Mollie is known under many names: Mollie Bond, Molly Van, Molly Vaughan, and even Mollie Whan — The Shooting of His Dear. For obvious reasons the author prefers the first of these titles.

*MOLLIE BOND*

*Come all you young men who handle a gun,*
*Be warned of shooting after the down sun.*
*A story I'll tell you; it happened of late,*
*Concerning Mollie Bond, whose beauty was great.*

*Mollie Bond was out walking, a shower came on;*
*She sat under a beech tree the showers to shun.*
*Jim Random was out hunting, ahunting in the dark;*
*He shot at his true love and missed not his mark.*

*With a white apron pinned round her he took her for a swan;*
*He shot and killed her; it was his Mollie Bond.*
*He ran to her; these words to her he said,*
*And a fountain of tears on her bosom he shed.*

*Saying 'Mollie, dear Mollie, you're the joy of my life.*
*I always intended to make you my wife.'*
*Jim ran to his uncle with his gun in his hand,*
*Saying, 'Uncle, dear uncle, I've killed Mollie Bond.'*

*Up stepped his dear uncle with locks all so grey;*
*Saying, 'Stay at home, Jimmie, and do not run away.*
*Stay in your own country till your trial comes on;*
*You shall not be molested, if it costs me my farm.'*

*The day of Jimmie's trial Mollie's ghost did appear,*
*Saying to the jury, 'Jim Random, come clear;*
*With my apron pinned around me he took me for a swan.*
*He shot and killed me and now I am gone.'*

**BUT THE AUTHOR** denies identity with the victim of the following cricketing mishap, which inspired Noel Petty to rewrite Henry Newbolt:-

*THERE'S A BREATHLESS HUSH.*

*There's a breathless hush in the Close tonight —*
*Ten to make and the match to win*
*As our number eleven squared up for the fight.*
*The first ball reared and grazed his chin,*
*And the second one jumped and split his thumb,*
*But little he cared for life's hard knocks*
*Till the third ball beat him and struck him plumb*
*In a place where wiser men wear a box.*

*He thought of his honour and thought of the School*
*And thought of the threat to his manly twitch,*
*And a voice inside said to him, 'Don't be a fool',*
*So 'Sod this!' he muttered, and limped from the pitch.*
*There's a breathless hush in the Close tonight —*
*Harrow won't play us again, they say.*
*'What bounder was that?' hissed the Head, death-white.*
*Matron blushed: 'It was Bond, sir, J.'*

**IN 1857 MR. T. BAKER** published *The Steam Engine: or, the Power of Flame, in Ten Cantos.* Canto X is particularly piteous with its threnody on the death of the Rt. Hon. William Huskisson at the opening of the Stockton to Darlington railway.

*The trains are stopp'd, the mighty Chiefs of Flame*
*To quench their thirst the crystal water claim;*
*While from their post the great in crowds alight,*

*When, by a line-train, in its hasty flight,*
*Through to striving to avoid it, Huskisson*
*By unforeseen mischance was over-run.*
*That stroke, alas, was death in shortest time;*
*Thus fell the great financier in his prime:*
*This fatal chance not only caused delay,*
*But damped the joy that erst had crown'd the day.*

HARRY GRAHAM'S *RUTHLESS RHYMES* contains many items of interest to those who study such matters. Here are three of them:-

*'There's been an accident!' they said,*
*'Your's servant's cut in half; he's dead!'*
*'Indeed!' said Mr. Jones, 'and please*
*Send me the half that's got the keys.'*

and

*Billy, in one of his nice new sashes,*
*Fell in the fire and was burnt to ashes;*
*Now, although the room grows chilly,*
*I haven't the heart to poke poor Billy.*

and

*Uncle, whose inventive brains*
*Kept evolving aeroplanes,*
*Fell from an enormous height*
*On my garden lawn, last night.*
*Flying is a fatal sport,*
*Uncle wrecked the tennis-court.*

IN DECEMBER 1934 there was a horrific boiler explosion at Powellton, West Virginia, on a railway belonging to a coal mining company. The engine itself landed on the first four coaches and the cab on a house. The following ballad was written by Charles A. Hudson.

*THE POWELLTON LABOR TRAIN EXPLOSION*

*It was a cold December morn,*
*The hour was about six o'clock,*
*There was a vast explosion*
*All Macdunn was filled with a shock.*

*Chorus*

*The explosion of the Powellton labor train,*
*On Armstrong Creek that day;*
*Leaving orphaned children and wives so dear,*
*Most to the poor house strayed.*

*The boiler of the labor train*
*Was hurled into the air,*
*And crashed down through the man-filled coach,*
*Burning, killing miners there.*

*The air was filled with death cries,*
*And was clouded with hot blinding steam;*
*They knew not what had happened*
*When they heard their comrades scream.*

*The fireman was dashed against the rocks,*
*The engineer was hurled into the air,*
*They found them in a tangled mass;*
*Blood chilling there.*

*It was an awful, horrible wreck,*
*One we shall never forget,*
*Those sixteen cold, perished men,*
*Made all eyes misty wet.*

*The hospital it was crowded,*
*The nurses and doctors had no rest;*
*They were soothing the blistered miners*
*They gave the live ones the best.*

*The lives of sixteen miners were taken*
*As I have told you before,*
*And many others wounded;*
*Forty-some, or more.*

*Those that were uninjured*
*On that unforgotten day,*
*Remained to help their workmates*
*Who in the wreckage lay.*

*The wreck scene was heartbreaking,*
*Crowds came from miles around,*
*To gaze upon those blood-smeared caps,*
*And love-letters that were found.*

*On the Twenty-seventh day of December,*
*The people's hearts were filled with cheer;*
*They knew not of their impending doom,*
*Bringing both misery and fear.*

*Neither do we know what day or hour,*
*Death may claim our soul,*
*Be kind to all you meet*
*As onward the years do roll.*

*Fathers, sons, husbands and sweethearts,*
*Your train might sometime explode,*
*So be prepared at any time*
*And stay on the righteous road.*

RUDYARD KIPLING in *The Secret of Machines* wrote these lines on the importance of treating machinery with respect. They deserve to be memorised by all who work with machines.

*We can pull and haul and push and drive,*
*We can print and plough and weave and heat and light,*
*We can run and race and swim and fly and drive,*
*We can see and hear and count and read and write...*

*But remember, please, the Law by which we live,*
*We are not built to comprehend a lie,*
*We can neither love nor pity nor forgive —*
*If you make a slip in handling us, you die.*

WE HAVE MENTIONED in a previous chapter the avalanche of snow at Lewes in 1836. As was the custom of the time, the local newspaper published some verses.

*The snow-storm came on a Christmas night,*
*And it pil'd its flakes on the cliff's broad height;*
*And there it lay in its fleecy pride,*
*When the cold sun gleam'd on the mountain's side.*

*And the cottagers dwelling beneath the hill,*
*Reckless of danger, regardless of ill,*
*Busied themselves in domestic care —*
*Mother and daughter and child were there.*

*The old man close to the fire-side stood,*
*To quicken the course of his torpid blood;*
*And the budding infant, with its glances sweet,*
*Gambol'd and crew at the old man's feet.*

*At length came a rushing sound,*
*And the avalanche made its fatal bound;*
*It dealt destruction to all beneath,*
*And whelm'd the inmates in darkness and death.*

*Sad and loud was the funeral wail*
*That was borne abroad on the biting gale;*
*They gather'd the victims no pow'r could save*
*And buried them all in one common grave.*

*The snow is melted, the storm is past,*
*And hush'd is the voice of the wintry blast;*
*Lightly and mild will the summer breeze blow,*
*And the Dead be forgotten who slept in the snow.*

Nothing like an avalanche for quickening the course of an old man's torpid blood.

HILAIRE BELLOC issued a warning, on this occasion to DIY enthusiasts:-

*Lord Finchley tried to mend the Electric Light*
*Himself. It struck him dead: And serve him right:*
*It is the business of the wealthy man*
*To give employment to the artisan.*

AN ANONYMOUS WRITER has commemorated the explosion that occurred in 1934 at the Gresford Colliery in Denbighshire, when 268 lives were lost:-

*THE GRESFORD DISASTER.*

*You've heard of the Gresford disaster,*
*The terrible price that was paid,*
*Two hundred and forty-two colliers were lost*
*And three men of a rescue brigade.*

*It occurred in the month of September,*
*At three in the morning, that pit*
*Was racked by a violent explosion*
*In the Dennis where gas lay so thick.*

*The gas in the Dennis deep section*
*Was packed there like snow in a drift,*
*And many a man had to leave the coal-face*
*Before he had worked out his shift.*

*A fortnight before the explosion,*
*To the shot-firer Tomlinson cried*
*'If you fire that shot we'll be all blown to hell!'*
*And no one can say that he lied.*

*The fireman's reports they are missing,*
*The records of forty-two days;*
*The colliery manager had them destroyed*
*To cover his criminal ways.*

*Down there in the dark they are lying,*
*They died for nine shillings a day.*
*They have worked out their shift and now they must lie*
*In the darkness until Judgement Day.*

*The Lord Mayor of London's collecting*
*To help both our children and wives,*
*The owners have sent some white lilies*
*To pay for the poor colliers' lives.*

*Farewell, our dear wives and our children,*
*Farewell, our old comrades as well.*
*Don't send your sons down the dark dreary pit,*
*They'll be damned like the sinners in hell.*

THE COALPORT FERRY TRAGEDY noted in chapter 2 was celebrated in 102 lines of turgid verse under the heading 'An accurate and poetical account of the DREADFUL CALAMITY which happened at Coalport, in the Parish of Madeley, Salop.' The opening lines give the flavour, and the author of this book will be happy to supply the remaining 90 lines to any reader with the stamina to get through them.

*Alas! Alas! the fated night,*
*Of cold October's twenty third,*
*In seventeen hundred ninety nine,*
*What cries! what lamentations heard!*
*The hour nine, when from yon pile,*
*Where fair porcelain takes her forms;*
*Where Energy with Genius joins*
*To robe her in those matchless charms;*
*A wearied band of artists rose,*
*Males and females old and young,*
*Their toils suspend to seek repose,*
*Their homes to gain they bent along.*

IN 1613 THE GLOBE THEATRE in London burned down. An anonymous balladeer issued a broadsheet:

*A SONNET UPON THE PITIFUL BURNING*
*OF THE GLOBE PLAYHOUSE IN LONDON*

*Now sit thee down, Melpomene,*
*Wrapped in a sea-coal robe,*
*And tell the doleful tragedy*
*That late was played at Globe;*
*For no man that can sing and say*
*Was scared on St. Peter's Day.*
*Oh sorrow, pitiful sorrow, and yet all this is true.*

*All you that please to understand,*
*Come listen to my story;*
*To see Death with his raking brand*
*'Mongst such an auditory;*
*Regarding neither Cardinal's might,*
*Nor yet the rugged face of Henry the eight.*
*Oh sorrow, etc.*

*This fearful fire began above,*
*A wonder strange and true,*
*And to the stage-house did remove,*
*As round as tailor's clew;*
*And burnt down both beam and snag,*
*And did not spare the silken flag.*
*Oh sorrow, etc.*

*Out run the knights, out run the lords,*
*And there was great ado;*
*Some lost their hats and some their swords,*
*Then out run Burbage too;*
*The reprobates, though drunk on Monday,*
*Prayed for the fool and Henry Condye.*
*Oh sorrow, etc.*

*The periwigs and drum-heads fry,*
*Like to a butter firkin;*
*A woeful burning did betide*
*To many a good buff jerkin.*
*Then with swollen eyes, like drunken Flemings,*
*Distressed stood old stuttering Hemings.*
*Oh sorrow, etc.*

*No shower his rain did there down force,*
*In all that sunshine weather,*
*To save that great renowned house,*
*Nor thou, O ale-house, neither.*
*Had it begun below,* sans doute,
*Their wives for fear...*
*Oh sorrow, etc.*

*Be warned, you stage strutters all,*
*Lest you again be catched,*
*And such a burning do befall*
*As to them whose house was thatched;*
*Forbear your whoring, breeding biles,*
*And lay up that expense for tiles.*
*Oh sorrow, etc.*

*Go draw you a petition,*
*And do you not abhor it,*
*And get, with low submission,*
*A license to beg for it*
*In churches,* sans *churchwardens' checks,*
*In Surrey and in Middlesex.*
*Oh sorrow, pitiful sorrow, and yet all this is true.*

THE CONVOLUTED ACCIDENT related in *The Sicknote* provides great scope for the imagination. It has also been set to music.

*THE SICKNOTE.*

*Dear Sir,*
*I write this note to you to tell you of my plight*
*And at the time of writing I am not a pretty sight.*
*My body is all black and blue, my face a deathly grey*
*And I write this note to say why Paddy's not at work today.*
*While working on the fourteenth floor, some bricks I had to clear.*
*Now to throw them down from such a height was not a good idea.*
*The foreman wasn't very pleased, he being an awkward sod.*
*He said I'd have to cart them down the ladder in my hod.*
*Now clearing all these bricks by hand it was so very slow,*
*So I hoisted up a barrel and secured the rope below*
*But in my haste to do the job I was too blind to see*
*That a barrel full of building bricks was heavier than me.*
*So when I untied the rope the barrel fell like lead*
*And clinging tightly to the rope I started up instead.*
*Well, I shot up like a rocket 'till to my dismay I found,*
*That half way up I met the bloody barrel coming down.*
*Well, the barrel broke my shoulder as to the ground it sped*
*And when I reached the top I banged the pulley with my head.*
*Well, I clung on tight 'though numbed with shock from this almightly blow.*
*And the barrel spilt out half the bricks fourteen floors below.*
*Now when these bricks had fallen from the barrel to the floor,*
*I then outweighed the barrel so started down once more.*
*Still clinging tightly to the rope I sped towards the ground*
*And I landed on the broken bricks that were all scattered round.*
*Well, I lay there groaning on the ground, I thought I'd passed the worst*
*When the barrel hit the pulley wheel and then the bottom burst.*
*And a shower of bricks rained down on me, I hadn't got a hope.*

*As I lay there moaning on the ground I let go the bloody rope.*
*The barrel being heavier, it started down once more*
*And landed right across me as I lay upon the floor.*
*But it broke three ribs and my left arm and I can only say*
*I hope you'll understand why Paddy's not at work today.*

**CONNOISSEURS OF BAD VERSE** will know the following West Indian ballad. It is found in the well known collection of Mr. D. B. Wyndham Lewis and is by James Grainger, MD, (1721–1767). It tells the melancholy story of Bryan (an English sailor) and Pereene ('the pride of Indian dames'). After a year's absence Bryan is so eager to return to the arms of his sweetheart that he cannot wait for the ship's boat to row him ashore but impetuously flings himself into the sea to swim to Pereene. We take up the story half-way through this tale of what Wyndham Lewis called the 'affecting parting not only of the lovers but of the lover himself.'

*In sea-green silk so neatly clad,*
*She there impatient stood;*
*The crew with wonder saw the lad*
*Repel the foaming flood.*

*Her hands a handkerchief display'd,*
*Which he at parting gave;*
*Well pleas'd, the token he survey'd,*
*And manlier beat the wave.*

*Her fair companions one and all*
*Rejoicing crowd the strand;*
*For now her lover swam in call,*
*And almost touch'd the land.*

*Then through the white surf did she haste*
*To clasp her lovely swain;*
*When, ah! a shark bit through his waist;*
*His heart's blood dy'd the main!*

*He shriek'd! his half sprung from the wave*
*Streaming with purple gore,*
*And soon it found a living grave,*
*And, ah! was seen no more.*

*Now haste, now haste, ye maids, I pray,*
*Fetch water from the spring:*
*She falls, she falls, she dies away,*
*And soon her knell they ring.*

*Now each May morning round her tomb,*
*Ye fair, fresh flow'rets strew,*
*So may your lovers 'scape his doom,*
*Her hapless fate 'scape you.*

**MEANWHILE, LUCASTA'S OLD MAN**, Colonel Richard Loveloss, now reduced in rank to that of a Captain of Industry, had been converted to loss prevention and decided to attend a course on the subject. He addressed his Board of Directors:-

*ON GOING TO THE COURSE*

*Tell me not, Sirs, I am insane*
*That from the cloistered hush*
*Of Board Room steeped in yearly gain*
*To safety course I rush.*

*True, new money must be spent;*
*A cash flow I'll require,*
*And with a safer plant prevent*
*A crash, a leak, a fire.*

*Yet cost effectiveness is such*
*As you must now explore;*
*For safety measures may cost much,*
*But compensation more.*

# CHAPTER 15

# APOCRYPHA

MANY OF THE ACCIDENT stories in this chapter may well have an element of truth in them, but as they have come to the author's attention in a less than authoritative version their credibility cannot be vouched for. Tales of mishap tend to get handed around, gathering the exotic detail and shedding the prosaic; nevertheless, even apocryphal accidents can make a point, if only to remind manufacturers that the unexpected can happen.

THE MAKERS of an early rotary drum washing machine advised in their instructions that, because of vibration, the washer should be mounted on a concrete base. A couple, who lived in an upstairs flat, were of that unusual breed who actually read the instructions supplied with appliances. They therefore cast a two foot thick concrete base for their new washing machine. The first time it was used the concrete base fell through the floor into the flat below.

It is one thing for English-speaking consumers to be faced with instructions written in muddy English. It is quite another problem when English is not used at all. The story is told of a woman who sent her young son to the supermarket to get a bottle of orange juice. He came back with a liquid that looked like orange juice and smelled of oranges. The label bore a picture of an orange but it was, however, written in German. The lady drank a tumbler of the liquid and became very ill. She had in fact drunk a large amount of orange-scented bubble bath.

EVEN DRINKING a glass of pure mineral water is not as safe as we may believe — that is, if we can credit an epitaph said to be at St. Giles Church in Cheltenham:-

*Here lies I and my three daughters,*
*Killed by drinking Cheltenham waters,*
*Had we'd kept to Epsom salts*
*We wouldn't be lying in these 'ere vaults.*

But perhaps the Hon. John Byng provides a clue to the source of its alleged impurity in his diary entry for 5th June 1781:-

'... Cheltenham-waters were first discovered by a grandmother of the present owner, Mrs. Skellingcoat, who observing the pigeons in great numbers frequenting the spring, tasted the waters and finding them mineral, drank of them herself, as well as several others, with great success; and then their merits were soon nois'd abroad.'

MANY OTHER EPITAPHS that have been quoted in humorous books are clearly spurious. Here is one such:-

*Here lies the body of Henry Bank,*
*Who struck a match to look in a tank.*
*They buried him quickly before he stank.*

and another:-

*Here lies the body of Martin Hyde,*
*Who fell down a midden and grievously died.*
*Here also his brother,*
*Who fell down another.*
*They now lie interred*
*Side by side.*

THE FOLLOWING INSCRIPTIONS, unlikely as they may seem, at least bear testimony to the care needed in the handling of firearms. The first is said to be at Woolwich.

'Sacred to the memory Major James Brush who was killed by the accidental discharge of a pistol by his orderly, 14th April 1831. Well done good and faithful servant'

And in Ulster a gravestone is alleged to have the following inscription:-

Erected to the Memory of
John Phillips
Accidently shot
As a mark of affection by his brother

SAILORS ARE KNOWN to be a superstitious lot. They are particularly fixed in their belief that Friday is an unlucky day, and a story consistently denied by the Admiralty runs as follows. To settle the matter once and for all, the naval authorities laid down the keel of a ship on a Friday, named her *Friday* and appointed a Captain Friday to command her. She made her maiden voyage on a Friday and sank with all hands.

But why spoil a good story by giving the facts. A recent report from the USA concerns an explosion at a disused ranger's station in one of the State Parks. The shock was felt by campers fourteen miles away. A local police detective looked into the affair and then explained why the concrete structure had so unaccountably blown up. It was all the fault of bats, he announced; a build-up of bat dung had produced methane gas which sank into the basement and was ignited by sparks from an electric sump pump. But alas for the plausibility of this rare piece of reasoning, as *The Chemical Engineer* pointed out, methane is lighter than air and could not therefore have sunk into the basement. Try again, Lieutenant Colombo.

TO SOUND THE ALARM on an industrial plant it is often necessary to telephone the control room. On one occasion a worker phoned the control room to warn them of a fire in part of the plant. The control room operator shouted down the telephone that he could not hear what the worker was saying because the alarm sounding on the control panel was making too much of a racket.

WHEN PAXTON'S CRYSTAL PALACE was built for the Great Exhibition in 1851 trolleys were devised to run along the guttering to carry the glaziers. It is related that one of the workmen had constantly to rebuke a young assistant for his wandering attention. No doubt the lad was unable to resist the wonderment of his bird's eye view of Hyde Park. But the glazier had other things to do and on one occasion impatiently clipped the boy around the ear for his inattention. In doing so he over-balanced and fell to the ground, bringing the young man with him. Amazingly it is said that the two between them suffered only one dislocated shoulder.

THE GLENFINNAN VIADUCT near Inverness is a beautiful structure, built at the end of the 19th Century. A firmly held belief in those parts relates to an incident during the construction works. Some of the pillars are hollow and before the superstructure to support the railway lines was built the tops of some of the columns were open. A horse and cart being led over planks resting across the top of a column fell into the interior and plunged some 100 ft down the hollow shaft. It proved too difficult to extract the carcass and it was buried inside the pillar.

Another entombment was an article of faith with all those who had any interest in the disaster-prone S.S. *Great Eastern.* As she approached New York on her third voyage in 1862 she struck an uncharted reef (since known as the Great Eastern Rock). She did not seem to be shipping water but because of her list it was suspected that water had got inside the double bottom of the port side. A diver was sent down to investigate and, horror-struck, reported back that he had heard a persistent tapping inside the outer shell of the ship. It is not known whether this was the cause or the effect of a story that a riveter had been trapped inside the double bottom. Although this tapping was later shown to have been caused by a loose cable shackle, belief in the interred riveter persisted.

In 1889 the *Great Eastern* was broken up for scrap and the public were still eager for the mystery to be cleared up. Their curiosity does not seem to have been satisfied; indeed it was further whetted on reports that the breakers had come across the skeletons of two men inside the ship's shell.

GARDENING ENTHUSIASTS tell this one. A tidy-minded gardener took a bucket of very hot water into his greenhouse to give it a good wash down. He dislodged a hanging basket which fell on the bucket, splashing him with hot water. He automatically jumped sideways, slipped up and gashed his ankle badly against the overturned galvanised iron bucket. The hanging basket, incidently, was planted with Achimenes (or Hot Water Plant) and Sanguinaria (Blood Root).

And gardening activities seem to have led to three family injuries. The story tells of a gardener who noticed that the summer growth of his wisteria was threatening his newly installed gutters. He set up his ladder, enroled his son to hold it steady, and climbed up with a heavy long-armed shrub pruner. All went well until he over-reached and felt himself toppling. To balance himself he dropped the pruner and clutched at the gutter. The pruner fell on his son who, startled and in pain, lurched against the ladder and dislodged it, bringing down father as well as the gutter he was clinging to. The son suffered a mild concussion and a crushed ear, but the father broke a leg in his fall. The story does not end here. The man's brother-in-law volunteered to replace the gutter but when lifting the ladder to set it up again he strained his back and was *hors de combat.* (A local handyman was eventually called in; he pruned the wisteria and replaced the guttering without mishap.)

IN 1987 AT GILBERTON, Alabama, a man is said to have used a cigarette lighter to search for his friend in the dark. The cigarette lighter caused a fire which killed the friend, engulfed two large oil storage tanks and forced the evacuation of most inhabitants of the town.

Also from the USA comes the tale of an inspector who was going through a works in Illinois when he observed a set screw projecting on a revolving shaft. He considered it particularly dangerous because the shaft was near a passageway and workmen were continually passing it. He drew the Manager's attention to it. 'Don't you think that this set screw had better be cut off before someone gets caught by it.' 'I don't think so' said the Manager, 'that set screw has been like that for years. No one has ever been hurt by it. The fact that it is exposed and can be seen by everyone makes it safe from causing an accident.'

The Manager had a mannerism when talking of gesticulating with his arms and he did so on this occasion. The sleeve of his coat then came into contact with the set screw and got caught. He was immediately dragged into the rotating shaft and whirled to death.

IN AUSTRALIA there is a statutory Workers Compensation scheme. One worker was awarded compensation when he dislocated his jaw while yawning at work. The employer was held responsible because of the boring nature of the work he required his staff to perform. Another compensation claim illustrates a knock-on effect of re-enacting an accident to find out what originally happened. A meat cutter at a Swiss hotel claimed for the loss of a finger in a piece of machinery. The insurance company's loss adjustor investigating the matter was a conscientious fellow and decided to operate the machine himself. He too had a finger cut off.

LAUREL AND HARDY ACCIDENTS such as this one are commonplace. A house proud manager was taking a walk around his plant to check

that all was in order when he noticed a pallet lying in the centre of a roadway. He was heard to say: 'Who left this pallet here?' In lifting it up to remove it he took a step forward and fell headlong into a dangerous hole.

THEN THERE WAS THE WORKER on a building site who was trying to saw a scaffolding plank in two using a hand saw. He was half-way through but found the going very hard. To make the job easier he balanced the scaffold plank on a drum, see-saw fashion. He sat on one end and persuaded the JCB driver to bring down the bucket smartly on the other end, the idea being that the plank would split on the fulcrum of the drum where the plank was partly sawn through. But the scaffold plank was stronger than the worker had bargained for, and when it was struck by the JCB bucket the poor workman was inevitably propelled upwards. He suffered considerable injury when landing back on the ground.

THE STORY OF A LITTLE OLD LADY who thought to dry her rain sodden cat in the microwave oven may be apocryphal, but the following tale seems probable. A young Australian golfer tried to dry a golf ball in a microwave oven. The ball exploded, broke his nose and deposited bits of rubber all over the kitchen.

LIGHTNING STRUCK a paint factory not so long ago, causing damage to many pieces of equipment and burning much of the paintwork. In one section of the plant the only paintwork not damaged was in a shade called Electric Blue. Elsewhere a fire broke out in the basement of an office block called Furnace House, while a sinuous dance routine by a Moscow State Circus 'snake girl' is said to have caused the spontaneous combustion of an Orthodox Priest's cassock.

THE TALE OF THE PARROT and the peer was found in a book of reminiscences published early this century. The peer was very fond of his parrot and went nowhere without it, even when travelling by train. He returned from a rail journey and an estate cart was backed onto the station platform towards the luggage it had come to collect. A helpful porter shouted 'Come on, come on' until the cart was suitably positioned. But atop the pile of suitcases and hampers was the parrot in its cage. The parrot, being an excellent mimic, picked up the cry and called for the driver of the cart to continue to 'Come on'. The driver did just that and backed the luggage, parrot cage and all, onto the railway line. Further details are lacking, but imagination can supply many endings to the tale. Did the cart also go over the platform edge onto the line? Was the parrot injured in its tumble? Did a through-train come roaring along? We shall never know.

THERE IS A TRADITION that the locomotive whistle was conceived after an accident in May 1833 at a level crossing on the Leicester to Swannington line. The line manager approached the Father of Railways, George Stephenson, about the practicability of equipping engines with a whistle that could be blown by the steam power, to give warning of a train's approach. The idea was accepted and the first 'steam trumpet' was fashioned by a musical instrument maker in Leicester.

IT IS ONLY TO BE EXPECTED that safety officers should be under particular scrutiny by the workforce for any infringement of the safety rules. And so it was at the site of a plastics factory which was being extended. The wearing of hard hats was a strictly enforced regulation, and when the safety officer — let us call him George — was seen to race bare-headed across the site the workmen downed tools to gaze at this extraordinary spectacle. They were not to know that the safety officer had received in his office a telephone message that the foreman's presence was immediately required at the maternity hospital where the birth of his first child was imminent. George, galvanised by this emergency, had unthinkingly left his office to speed the foreman on his way. It was therefore unfortunate that, just as George passed under it, a piece of scaffolding should have fallen, accidentally dropped no doubt by one of the incredulous workmen watching his safety officer do the unthinkable.

George, badly concussed and with severe lacerations to his face, was in hospital for several weeks. Another version of this story has it that George himself was the father-to-be and was taking a short cut across the site to the car park. But whatever the truth (if any), poor George was seen by all to have broken one of his own rules.

COCKROACHES ARE notorious long-livers and the lady's dilemma in the following incident is understandable. She was in the bathroom at a hotel when she saw one. She stamped on it and scooped it into the lavatory

bowl. It refused to die, and she used a large amount of hair lacquer from her can of hair spray before the creature finally gave up and was flushed away. Shortly afterwards her husband came in to use the lavatory and while sitting threw a cigarette end into the pan. A small explosion, caused by the ignition of the butane gas in the hair spray propellant, seriously burned the gentleman where he was most sensitive. The two ambulance men who came to take him to the casualty department laughed so much that they dropped the stretcher bearing the poor man, who thereupon added a broken pelvis and some broken ribs to his original injury.

IN JANUARY 1838 the Royal Exchange in London burned down, and it was unfortunate that the bitter winter cold had frozen much of the water that would otherwise have been immediately available to douse the flames. The building had a grand clock tower with a carillon that played twelve tunes, one for each hour of the day. The old Scottish song 'There's nae luck about this house' was one of the tunes, and contemporary writers maintained that when the heat of the fire caused the mechanism of the carillon to jam this tune continued playing until the clock tower collapsed into the inferno.

AND NOW FOR SOME flying stories. They are all so improbable that it is hard to believe they weren't just invented as entertaining dinner party tales.

For the first one we have to go back to the nineteen-thirties for a grotesque accident that happened to five German glider pilots. They were carried into a thunder cloud above a mountain range and decided to bail out of their aircraft. With their parachutes they were carried by strong upcurrents into an atmosphere of super-cooled vapour and were covered by layers of ice until they finally fell as gigantic hailstones. Only one of these pilots survived.

This next story was frequently told to flying cadets who aspired to a career in the RAF. A Wellington bomber had made a bad landing and when it came to a halt the instructor got out to discover the reason, leaving the engines just ticking over. He inspected the undercarriage from the rear; he inspected it from the side. He went round to the front for a careful examination and then walked slowly backwards to check the alignment. Only when he noticed the propeller immediately in front of him did it dawn on him that he had walked backwards straight through the slowly rotating propeller.

It was the pilot's nerve that was at risk in our last flying story. In the days when Aden was a British Protectorate a Colonial officer had died up country and a light aircraft was sent to bring his body back to the port for onward shipment to England for burial. The pilot of the tiny aircraft was a fey young man (said to be the seventh child of a seventh child), and we can without much difficulty imagine his apprehension when, hearing a creaking behind him while at high altitude, he turned his head to see the coffin lid slowly lifting. Such are the hazards of carrying, or being carried in, a temporary coffin in an unpressurised aircraft.

FINALLY THE STORY of the worker in a sawmill. He had a slight accident one day and did not realise that he had lost the first two fingers of his right hand until he clocked off early and attempted to make a parting gesture to his foreman....

# CHAPTER 16

# QUOTES

IT IS NOT KNOWN who first pointed out that the light at the end of the tunnel may be the headlamp of an oncoming train, nor do we know about whom it was said he was so boring that he invented clichés. Here are some wise sayings, old saws, new saws, mottoes, maxims and sundry quotations that the reader might find useful. There is no particular order but pride of place must go to William Shakespeare, the most quoted writer of all:-

*Out of this nettle, danger, we pluck this flower, safety.*
King Henry IV, Part 1.

*A little fire is quickly trodden out,*
*Which, being suffer'd, rivers cannot quench.*
King Henry IV, Part 3.

*There is a history in all men's lives,*
*Figuring the nature of the times deceas'd;*
*The which observ'd, a man may prophesy,*
*With a near aim, of the main chance of things*
*As yet not come to life, which in their seeds*
*And weak beginnings lie intreasured.*
King Henry IV, Part 2.

WHERE NO ATTRIBUTION IS GIVEN, the originator must be assumed to be untraceable. Some are given in the original language because they seem to be more profound that way.

Now and then there is a person born who is so unlucky that he runs into accidents which started out to happen to somebody else.
Don Marquis, *Archy's Life of Mehitabel.*

*****

An injury prevented is a benefaction, an injury compensated — an apology.
Quoted by H. W. Heinrich in *Industrial Accident Prevention* but author unknown.

*Pars tutior sequenda est.* (The safer course must be followed)

*****

It is folly to expect men to do all that they may reasonably be expected to do.

Richard Whately, Archbishop of Dublin (1787–1863)

*****

The chapter of accidents is the longest chapter in the book.

John Wilkes (1727–1797)

*****

Experience teaches slowly, and at the cost of mistakes.

J. A. Froude (1818–1894)

*****

There is perhaps a feeling in Scotland that, as a result of the restrictions imposed, whisky is already so sadly diluted that no risk should be taken of any fortuitous aggravation of this treatment.

On the desirability of sprinklers in certain industries, *Report of the Royal Commission on Fire and Fire Protection, 1923.*

*****

Better one safe way than a hundred on which you cannot reckon.

Aesop, *The Fox and the Cat.*

*****

The saddest part of this matter is that no lesson of any kind has been taught by the event, as everyone who has studied the subject either theoretically or practically knew beyond any possibility of a doubt what the whole action of the fire and smoke would be under such circumstances; and moreover, the lessons and warning of recent years had prepared all concerned for the terrible catastrophe precisely as it actually occurred.

Captain E. M. Shaw, Chief Officer of the London Metropolitan Fire Brigade, in his report on the fire at the Theatre Royal, Exeter, in 1887.

*****

Pish! A woman might piss it out.

Sir Thomas Bludworth, Lord Mayor of London, dismissing a fire that had broken out at a bakery in Pudding Lane in the small hours of Sunday, 2 September 1666.

Casualties in aviation are generally caused by involuntary vertical movement towards the ground.

From the prospectus issued to potential shareholders in a fleet of helium and hot-air powered airships.

*****

Man seeketh in society comfort, use and protection.

Francis Bacon (1561–1626)

*****

Constant disappointment has as little effect upon an English politician as upon an alchemist.

Robert Southey (1774–1843)

*****

This place is a death trap.

Union spokesman, on the headquarters of the Health and Safety Executive during renovations, when the staff were on strike because the building was 'unsafe'.

*****

A man gazing at the stars is proverbially at the mercy of the puddle in the road.

Alexander Smith (1830–1867), *Dreamthorp*.

or as Edmund Spenser (1552?–1599) put it:-

And he that strives to touch the stars
Oft stumples at a straw.

*****

Shall we call in the experts or foul it up ourselves?
Banner in the room of a chemical works used by the Youth Training Scheme.

*****

Reorganisation is a splendid method of producing the illusion of progress whilst creating confusion, inefficiency and demoralisation.
Petronius Arbiter, 60AD.

*****

He that will not apply new remedies must expect new evils; for time is the great innovator.
Francis Bacon (1561–1626)

*****

'Why do they call it 'LESSONS' ?' asked Piglet. Pooh thought for a long time. 'I suppose,' he said, 'I suppose it is because lessons are what you learn to lessen the chance of the same thing happening again.'
(with apologies to A. A. Milne)

*****

*Quidquid agas, prudenter agas, et respice finem.*
(Whatever you do, do cautiously, and look to the end.)

*****

What we anticipate seldom occurs; what we least expect generally happens.
Benjamin ('Murphy') Disraeli (1804–1881)

*****

There are some know-it-alls who keep count of all disasters — Chernobyl and the *Admiral Nakhimov*, the loss of the atomic submarine, the coal mine explosion — and see them as proof that if anything has changed in our country it is for the worse. No, it is not that there are more disasters, but that there is more honest and frank information about them and, had there been such openess

before, there could be less negligence and fewer disasters today.
Academician Vitaliy Goldanskiy, *New Times*, Moscow, 1987.

*****

People only do two sorts of things. Those things they want to do and those things the boss checks up on.
John Garnett, Director, The Industrial Society.

*****

The errors of a wise man make your rule,
Rather than the perfections of a fool.
William Blake (1757–1827), *On Art and Artists.*

*****

The difference between a pessimist and an optimist is that a pessimist has more information.

*****

Misfortunes are often accidents, yet the calamities inflicted on us by the hand of God are very few in proportion to those which come from our own errors.
William Lamb, 2nd Viscount Melbourne (1779–1848)

*****

Everything that can be invented has been invented.
The Director of the United States of America
Patent office, 1899.

*****

When eating an elephant take one bite at a time.
General Creighton W. Abrahams.

Study the past if you would divine the future.
Confucius.

*****

From bad judgement comes experience, from experience comes good judgment.
Author unknown.

*****

*Mortui Vilpos Docent.* (The Dead Teach the Living.)

*****

A non-return valve is an invention of the devil, put on earth to torment man. It functions often enough to show great promise, but always lets you down when you really need it.

*****

In every affair consider what precedes and follows and then undertake it.
Epictetus, 1st century BC.

*****

Had I have known, I wouldn't have agreed.
Comment made by a senior manager after the fire which followed his instructions to a plant manager to operate a reactor inside the flammable region.

*****

This horse has diabetes.
Laboratory report on a sample of water from the swimming pool of a London club sent for analysis after complaints by members.

*****

Any fool can build a bridge but it takes an engineer to make it just not fall down.
*14 Steps of Quality.*

*****

Knowledge is of two kinds. We know a subject ourselves, or we know where we can find information upon it.
Dr. Samuel Johnson (1709–1784)

*Quid quisque vitet, nunquam homini satis*
*Cautum est in horas.*
(Who can hope to be safe? Who sufficiently cautious?
Guard himself as he may, every moment is an ambush.)
Horace, *Odes*.

*****

We used to ask people to take more care and act safely. If there was an accident it would be investigated. Now we have a management system for it all.

*****

He is truly wise who gains from another's mishap.
Publilius Syrus, Maxim 825 (c.43BC)

Another of whose maxims lays it down that:-

He is safe from danger who is on guard even when safe.

*****

*Nous avon tous assez de force pour supporter les maux d'autrui.*
(We have all enough strength to bear the misfortunes of others.)
Duc de la Rochefoucault (1613–1680)

*****

It should not be necessary for each generation to rediscover principles of process safety which the generation before discovered. We must learn from the experience of others rather than learn the hard way. We must pass on to the next generation a record of what we have learned.
Jesse C. Ducommun, Vice President, American Oil Company.

*****

.... nothing [is] so instructive to the younger Members of the Profession, as records of accidents in large works, and the means employed in repairing the damage. A faithfull account of those accidents, and of the means by which the consequences were met, [is] really more valuable than a description of the most successful works.
Robert Stephenson, Victorian Engineer, 1886. Son of George Stephenson.

The villain that brought upon the earth so injurious a thing does not deserve to have his name remain in the memory of men.

On the question of who invented gunpowder,
Sebastin Miller, *Cosmographie*, 1584.

*****

It is confidence which causes accidents and worry which prevents them.

Professor J. E. Gordon

*****

'High pressure steam boilers would not scatter death and destruction around them if the dishonesty of avarice did not tempt their employment, where the more costly low pressure would ensure absolute safety.'

The Rev. Dr. Opimiam in *Gryll Grange* by Thomas Love Peacock, published 1860.

*****

This question of danger should be looked at in just an ordinary commonsense way; no heroics, or anything of that kind. Let me put it like this. Take any ordinary sort of man leading a life of humdrum routine. During the whole of it he runs a certain amount of risk, doesn't he? He runs the risk of being knocked down by a motor-bus or a motorcar; of having a tree fall on him in a gale; of being in some railway accident; of catching some complaint that may kill him. But all those risks are stretched over a number of years. As for me, all I'm doing on an Atlantic flight is to cram a life-time of risk into a period of say thirty-six hours. I am taking all my medicine in one dose, you may say.

Colonel Dan Minchin (Daredevil Dan). He disappeared on his next Atlantic flight.

*****

A wise man learns from his own experience, but a wiser man learns from the experience of others.

Author unknown

*****

Learn from the mistakes of others, you'll never live long enough to make them all yourself.

An old pilot's saying.

A superior pilot is one who uses his superior judgement to keep himself out of situations that might require excessive use of his superior skill.

Another old pilot's saying.

*****

No jesting with edge tools, or with bell ropes.

An old Sussex proverb.

*****

By this time the report of the accident had spread among the workmen and boatmen about the Cobb, and many were collected near them, to be useful if wanted, at any rate to enjoy the sight of a dead young lady, nay, two dead young ladies, for it proved twice as fine as the first report.

Jane Austen (1775–1817), *Persuasion*.

*****

Have more strings to thy bow than one; it is safe riding at two anchors.

John Lyly, *Euphues* (1579)

*****

Dangers by being despised grow great.

Edmund Burke

*****

Unlike all other organisms Homo Sapiens adapts not through modification of its gene pool to accommodate the environment but by manipulating the environment to accommodate the gene pool.

M. T. Smith and R. Layton, *The Sciences*, 1989.

*****

When in doubt — mumble; when in trouble — delegate; when in charge — ponder.

James H. Boren.

*****

A man may surely be allowed to take a glass of wine by his own fireside.

Richard Brinsley Sheridan, to a friend who expressed surprise that he should be taking refreshment in a nearby coffee house while his theatre in Drury Lane was burning down in 1809.

*****

To a man who has an instinctively mechanical mind — and no other can be an engineer — the advice I have given you above is all I need say; but this advice is the result of a good deal of experience, purchased by failures of my own, and by looking at those of others...

Isambard Kingdom Brunel, in a letter to one of his assistants dated December 30, 1854.

*****

Nature and Nature's laws lay hid in night:
God said, *Let Newton be*! and all was light.

Alexander Pope (1688–1744)

But not for long the devil howling Ho!
Let Einstein go and restore the status quo.

Author unknown

What would life be without arithmatic, but a scene of horrors.
Rev. Sydney Smith (1771–1845)

*****

A man should never be ashamed to own he has been in the wrong, which is but saying, in other words, that he is wiser today than he was yesterday.
Jonathan Swift (1667–1745)

*****

The first essential in the treatment of burns is that the patient should be removed from the fire.
First Aid Manual.

*****

The vulnerability of chemists to high levels of chemical exposure means that they are often regarded as prime targets for disease. However, according to the results of a 25-year survey carried out by the Society, professional chemists live longer than other workers in the same social groups.
*Chemistry in Britain*, August 1993.
Journal of the Royal Society of Chemistry.

MURPHY IS ALMOST as well known a lawgiver as Moses. We offer a selection of developments of his First Law:-

When everything is going well, disaster is about to strike.

A dropped tool will always land where it will do the most damage (also known as the Law of Selective Gravitation).

All pieces of equipment have an innate dislike of people.

The more innocuous a design change appears, the further its influence will extend.

If anything that should have gone wrong does not go wrong, it would, in the long run, have been better if it had gone wrong.

and anon has pointed out that the difference between the laws of nature and Murphy's Law is that the laws of nature ensure that it goes wrong in the same way every time.

# CHAPTER 17

# THE ONE THAT GOT AWAY

EDMUND SPENSER'S SWEET THAMES that ran softly was known throughout Europe in 1858 as the Great Stink. Hardly surprising. In the words of one appalled peer in a House of Lords debate: 'Eighty-two million gallons of pure water are taken out of the Thames every day, and before they are poured back into the river they have to pass through all the water closets in London.'

The purity of the water taken out of the Thames was highly questionable, considering that breweries, soap works, glue factories, knackers' yards, slaughter houses, tanning works and other riverside factories poured their toxic waste unchecked into the river. Most of London's 370-odd sewers were open along their whole length; all except one were open at their outfall when the tide was ebbed. The exception was the newest and biggest, the Victoria sewer.

The sickly inhabitants of London had put up with a putrid river for many years, but in the hot, rainless summer of 1858 the Houses of Parliament found the Great Stink altogether overpowering and a Select Committee was appointed. The windows of Parliament were hung with curtains soaked in chloride of lime and chloride of zinc, while the Select Committee considered the practicality of pouring barrow-loads of slaked lime into the sewers at the point where they disgorged their contents into the river. They studied the results of a dainty little experiment conducted by the revered Professor Faraday. They resigned themselves to the reproaches of the press. They listened to Mr. Goldsworthy Gurney.

MR. GURNEY, a Cornish inventor and charlatan, took it into his head that the way to purify London was by burning off the poisonous gases at a height where they would do no harm. He had a pipe built from the Victoria Sewer through New Palace Yard and up into the Clock Tower. The gases had now been brought high above the Thames and Mr. Gurney put a lighted match to the end of the pipe. It went out. He then joined a coal stove to the pipe in the Clock Tower and lit it, and with great self-satisfaction informed the Select Committee that the gases were now being burned off.

What Mr. Gurney did not know, but what engineer Joseph Bazalgette discovered, was that the pipe to the Clock Tower from the sewer had always been blocked by a trap-door (non-return valve), originally fitted as a safety device, which had either fallen or been pushed into place. Bazalgette also discovered that the Victoria Sewer was full of an explosive mixture of sulphuretted hydrogen from the sewer and coal gas from a broken gas main. Had Mr. Gurney's pipe not been blocked by the trap-door there is little doubt that his first lighted match would have caused an explosion which would have blown up the Clock Tower, much of the Palace of Westminister, and a great length of the Victoria Sewer. Needless to say Bazalgette had the trap re-sealed, although not before enough of the gas had escaped into the Clock Tower to cause a small explosion the next time Mr. Gurney's stove was lit.

Yet again the big one had been avoided. Guy Fawkes had failed and Mr. Gurney's misguided effort had been a near miss. No more was heard of the plan to lead the sewer gases skywards and burn them off. Instead, a Bill was rushed through Parliament allowing slaked lime to be dumped into the Thames, the weather cooled, the rains came, the river sludge was diluted to its normal fetid state, and mortality figures for dysentery, cholera and other fevers began to drop. The Great Stink was ended, and when Parliament was prorogued Queen Victoria sent a message:-

> 'My Lords and Gentlemen,
>
> The sanitary condition of the Metropolis must always be a Subject of deep interest to Her Majesty, and Her Majesty has readily sanctioned the Act which you have passed for the purification of that noble River, the present state of which is little creditable to a great Country and seriously prejudicial to the Health and Comfort of the Inhabitants of the Metropolis...'

# CHAPTER 18

# JANUS

IT MAY SEEM to some readers that the recounting of all these accidents of the past is not relevant to the hazards of today. But think about it for a moment. In chapter 2 is an account of the accident to Samuel Wood who lost his arm and shoulder blade, followed by a similar accident fifty years later and a fatal one a couple of years after that. These are but a few of the appalling accidents of earlier times, before the principle of guarding machinery had sunk into the industrial psyche. But there is no room for complacency — limbs and lives are still being lost and Kipling's lines about machinery (chapter 14) are as applicable today as when he wrote them.

IN HIS RECENT BOOK *Lessons from Disaster*, Trevor Kletz states:-

> 'It might seem to an outsider that industrial accidents occur because we do not know how to prevent them. In fact, they occur because we do not use the knowledge that is available. Organisations do not learn from the past or, rather, individuals learn but they leave the organisation, taking their knowledge with them, and the organisation as a whole forgets.'

How can we learn and pass on lessons that plainly need to be learned and passed on. We can read the many books that have been written about accidents, but do we really take on board the knowledge we acquire from them? The professional can use accident databases on computer, but these seldom include the lessons they should teach. Despite the efforts of many, some organisations still do not share the lessons they have learned from their own accidents and indeed are reluctant to record them. It has even been reported that the results of an investigation into a fatal accident in one organisation were contained in a single paragraph. Is this any different from the bald entry for Henry Wells in the parish register quoted at the end of chapter 8?

I have tried to use humour, the unusual event and some illustrations to help us along the learning curve. The incident at Marton-in-Cleveland in

1812 (chapter 11) where three men were removed dead from a well reminds us of the necessity to test for oxygen and to use a safety harness when entering a confined space. Even the French girl who used her own hair as a kind of safety rope and the workmen at Beziers who used a cock and a cat to test the atmosphere can be enlisted in the process because accidents of this nature are still happening around the world. In 1978 a man collapsed when cleaning out a tank on a marine chemical tanker in the Pacific. A fellow worker went to his rescue, followed by another and another and another and another and another. Seven men died in that tank, asphyxiated by oxygen deficiency.

The explosion caused by the old crone knocking out her pipe in the presence of gunpowder (chapter 5) had far more gruesome results than we have included in our quotation from the press report, and we may be tempted to write this accident off to the ignorance of one elderly and uneducated woman. But look again at the item in chapter 10 where a man lit a cigarette in a tank he must have known was filled with dangerous fumes — why else did he think he had been provided with a breathing apparatus?

ALTHOUGH IN SOME CASES it may appear to be the action of an individual that precipitated an accident, it is wise to look for possible underlying causes to ensure there is no repetition of the action. Take the example of somebody opening the wrong valve and setting off a train of mayhem. Was he (or she!) sufficiently well trained to know one valve from another? Were the 'right' and 'wrong' valves clearly differentiated? Was the operative overtired from working too long hours? Was the design of the plant substandard in not incorporating a belt-and-braces mechanism so that even if a wrong choice of valve were made there would be a safety back-up? Even if the operative was shown to have celebrated his birthday too well during his lunch hour, did the company have an unambiguous and *known* policy about drinking alcohol during working hours?

A No Drinking policy, of course, is of little value unless it applies at all levels of a company. Leaving aside the virtue of leading by example, a decision maker plainly needs as clear a head as a plant hand operating complex machinery. The tragedy of 1883 in Sunderland when 186 children were crushed to death is an example of decisions being made without thinking through the possible consequences. The decision to lock one of a pair of swing doors was basically an alteration to the design of the theatre, and it is nowadays a commonplace (or should be) that any alteration to a design must be assessed for hazards that might be created as a result of the alteration. This failure on the part of the theatre management was compounded by allowing the performers to throw the presents only into the main body of the hall without insisting that one of the performers go into the gallery to distribute presents to the children upstairs. It all seems obvious to us today, but there is never room for smugness where safety is concerned. Hindsight is no substitute for foresight.

ARE THERE ANY NEW HAZARDS around today? There are a few but too frequently the hazards that caused the accidents of yesterday are the same ones that cause the accidents of today. After all, a badly designed and installed gas heater will cause the same carbon monoxide poisoning as a pan of burning charcoal, and we have temporary spectator stands today that collapse as disastrously as they did when old Lord Lovat was beheaded.

Safety is not an intellectual exercise designed to keep a number of experts in employment. It is to do with people. The man or woman who is killed or injured in an accident is not an isolated statistic, the subject of a case study. Spouses, children, parents, brothers, sisters, friends — all are affected. Many believe it would have a salutory effect on a company's safety policy if the Directors were the ones who had to call on a family to break the news that their breadwinner has been totally incapacitated. It may be no coincidence that the multinational chemical company DuPont, with a safety record that is the envy of all others, lost one of its directors, Alexis I. du Pont, in the late 19th century through an industrial explosion. Alexis found his clothes on fire and jumped into a tank of water (fortunately not covered in ice) while the building around him disintegrated in an explosion. He went to the assistance of other workers in another building but this also exploded and he was killed.

Risk and hazard identification are concepts that can be taught to primary school children. The Health and Safety Executive has supported the introduction of an *Activity Box* to help primary school children understand how these concepts apply in the world around them. Adults too have an activity box; we call it a brain, and we must use it to learn how to identify the hazards of life so that accidents can be avoided. At every opportunity we must identify the hazards and assess the risks of causing injury to ourselves and others.

IN THE 18TH CENTURY Horace Walpole's aunt thought it was up to Parliament to abolish accidents, and there are many around today who suffer from this Aorta Syndrome ('they oughta do something'). Safety legislation has come a long way since the Code of Hammurabi but there is a limit to what it can achieve. It can oblige manufacturers to put appropriate warnings on labels but it cannot compel the buying public to read the labels, let alone heed the warnings. Laws can take a long time to get enacted; they have to be policed; they can be limited in scope, as were the 19th century regulations covering safety at places of public entertainment which did not apply to private clubs — we noted in chapter 9 the consequences of this shortsighted legislation when sixteen people lost their lives at the dramatic club in Spitalfields.

Moreover, as Charles Dickens reminds us (see Chapter 2) there are frequently lobbyists around trying to frustrate the intentions of government. Another writer, Robert Southey, summed the situation up in *Letters from England* in 1807:-

> 'If the elixir of life were actually to be discovered, the furnishers of funerals would present a petition to Parliament praying that it might be prohibited, in consideration of the injury they must otherwise sustain; and, in all probability, Parliament would admit their plea.'

Legislation does indeed have a star role to play but it is a finite role and it really will not do to rely solely on the law. Those responsible for the safety of others, whether parents of small children or employers of a vast workforce, cannot escape moral culpability just because Parliament has not banned this, compelled that, or set up a quango to supervise the other. There is nothing quite like self-regulation.

THOSE IN CHARGE OF COMPANIES and other organisations have a special responsibility, of course, because their decisions affect the lives of so many more people. Even when no deaths or injuries are involved it is no use just shrugging our shoulders when a company has to report diminished profits because of lost production time, damaged equipment or compensation for harm to the environment; the company may find itself forced to lay off men and women or retire them early, with all the distress that this will cause. Trainee financial auditors are always told that their first duty is to the little old lady with her life savings in a company — she too will feel the effects of a company's loss of profit caused by a bad safety record.

JANUS WAS A GOD of the ancient Romans who is depicted as having two faces, one looking backwards and the other to the front. He was a guardian of beginnings and the month of January is named after him because he looked back to the past year and forward to the year to come.

Having read about some of the hazards that confront the human race, the Janus approach to safety may suggest itself to us: looking to the past to learn what has happened and to the future to identify what could happen. The necessary precautions then become a matter of common sense.

# COPYRIGHT ACKNOWLEDGMENTS

Considerable effort has been taken to trace the copyright holder of various items in this book. If I have inadvertently not obtained the appropriate permission I apologise, but I hope that the owner has enjoyed the book and appreciates the use of the material. I wish to acknowledge the following for permission to reproduce some of the stories given in this book.

Barrie and Jenkins Ltd for extracts from *The History of Tunnelling* by Gosta E. Sanderstom.

Chapman and Hall Ltd for extracts from *Michael Faraday — A Biography* by L. P. Williams.

David and Charles Publishers for the item from *The Railway Navvy.*

Elizabeth Hamilton for the story on lightning at Eton College.

Felling Male Voice Choir for the words of the Felling Disaster.

HarperCollins Publishers Ltd. for quotations from the *Father of the Submarine* by W. S. Murphy, *Curiosities from Parliament* by Stanley Hyland and *The Best of Tombstone Humour* by Richard De'ath.

Hazardous Cargo Bulletin for the story on WATER and other anecdotes.

Martin Secker and Warburg Ltd. for extracts from *Aubrey's Brief Lives* edited by Oliver Lawson Dick.

Nöel Petty for the poem *There's a Breathless Hush.*

Penguin Books for extracts from the *Chronicles of the Crusades.*

Peters, Fraser & Dunlop Group Ltd. for reprinting *George who played with a Dangerous Toy* by Hilaire Belloc.

Post Office Archives for the items on the Mail Coach Service.

*Private Eye* for the story on the safety exercise on a plant.

*Scottish Epitaphs* by Raymond Lamont-Brown for the epitaph in the Durness Churchyard and published by W & R Chambers.

Sussex Record Society for references to medieval inquests.

University of Wales, Swansea, for quotations from the letters of Robert Morris.

Vallentine Mitchell Publishers for extracts from *Marie Antoinette* by Andre Castelot, translated by Denise Folliot.

Whitegate Primary School for the reference to conkers collected in the First World War.

# BIBLIOGRAPHY


Altman, L. K. 1987 *Who Goes First* Random House

Atlas Assurance Company 1958 *On Risk* Eyre and Spottiswoode Ltd.

Assheton, R. 1930 *History of Explosions on which the American Table of Distances was Based* Institute of Makers of Explosives

Biasutti, G. S. *History of Accidents in the Explosives Industry* Private publication

Blackstone, G. V. 1957 *A History of the British Fire Service* Routledge & Kegan Paul

Blake, Mrs. Warrenne 1911 *An Irish Beauty of the Regency* John Lane

Brock, A. St. H. 1949 *A History of Fireworks* Harrap

Brooke, D. 1983 *The Railway Navvy* David and Charles

Buckingham, Duke of 1856 *Memoirs of the Court of England during the Regency* Hurst and Blackett

Bushell, P. 1983 *London's Secret History* Constable

Byng, J. 1954 *The Torrington Diaries* Eyre and Spottiswoode

Castlelot, A. 1957 *Marie Antoinette* Translated by D. Folliot, Vallentine and Mitchell

Caulfield, J. 1820 *Remarkable Persons* Volume 4, H. R. Young

Cecil, D. 1954 *Lord M* Constable

Chaloner, W. H. 1959 *Vulcan — The History of One Hundred Years of Engineering and Insurance* Vulcan Boiler and General Insurance Co. Ltd.

Chorlton, P. 1988 *Cover Up* Grapevine

Collier, Basil. 1974 *The Airship* Hart-Davis MacGibbon

Connell, B. 1957 *Portrait of a Whig Peer* Andre Deutsch

Curzon, R. 1849 *Visit to the Monasteries in the Levant* John Murray

De'ath, R. 1987 *The Best of Tombstone Humour* Javelin Books

Dick, Oliver Lawson (editor) 1949 *Aubrey's Brief Lives* Secker and Warburg

Dickinson, H. W. and Titley, A. 1934 *Richard Trevithick. The Engineer and the Man* Cambridge University Press

Douglas, D. C. (editor) 1979 *English Historical Documents* Eyre and Methuen

Dumpleton, B. and Miller, M. 1974 *Brunel's Three Ships* Venton

Egremont, Lord, 1968 *Wyndham and Children First* Macmillan

Elam, J. F. 1986 *St. Mary's Church East Bergholt* East Bergholt Church Council

Emmerson, G. S. 1981 *The Greatest Iron Ship* David and Charles
Ernst, W. undated *The Life of Lord Chesterfield* Swan Sonnenschein
*Faber Book of Popular Verse* 1971
Falk, B. 1944 *The Berkeleys of Berkeley Square* Hutchinson

Gaussen, A. C. C. 1904 *A Later Pepys* John Lane
Glaister, J. and Logan, D. D. 1914 *Gas Poisoning in Mining and Other Industries* E. & S. Livingstone
Gleick, J. 1988 *Chaos* Heinemann
Gould, S. J. 1979 *Ever Since Darwin* Norton
Gronov, Capt. 1900 *Reminiscences and Recollections 1810–1860* John C. Nimmo

Hague, D. B. and Christie, R. 1975 *Lighthouses* Gower Press
Hamilton, E. 1965 *The Mordaunts: An Eighteenth Century Family* Heinemann
Hartley, D. 1978 *Water in England* Macdonald & Jane's
Herbert, Lord (editor) 1939 *The Pembroke Letters* Johnathan Cape
Hewison, C. H. 1983 *Locomotive Boiler Explosions* David and Charles
Hyland, S. 1955 *Curiosities from Parliament* Allan Wingate

Jennings, Humphrey 1985 *Pandaemonium* Andre Deutsch
Joby, R. S. 1917 *The Railway Builders* David and Charles
Joinville 1963 *Life of Saint Louis* Translated by M. R. B. Shaw, Penguin Classics
*Journal of the Royal Institute of Chemistry*
*Journal of the Royal Society of Medicine*

Kletz, T. 1988 *Learning from Accidents in Industry* Butterworths
Kletz, T. 1993 *Lessons from Disasters* Institution of Chemical Engineers
Knighton, Lady 1838 *Memoirs of Sir William Knighton* Carey, Lea & Blanchard

Lamont-Brown, R. 1990 *Scottish Epitaphs* Chambers
Latham, F. A. 1993 *Vale Royal* The Local History Group
Liddell Hart, Capt. B. H. (editor) *The Letters of Private Wheeler 1809–1828* Cedric Chivers Ltd.
Llanover, Lady (editor) 1862 *Autobiography and Correspondence of Mary Granville* Richard Bentley
Loaring, H. J. 1873 *Epitaphs — Quaint, Curious and Elegant* William Tegg
Lyle, K. L. 1985 *Scalded to Death by Steam* W. H. Allan

Marshall, V. C. 1988 *The Boston, Mass., Incident of January 15th 1919* Loss Prevention Bulletin 082
Maxwell, Sir H. (editor) 1904 *A Selection from the Correspondence & Diaries of the late Thomas Creevey, M.P.* John Murray
Morozzo, Count 1795 *The Repertory of Arts and Manufacturers: Consisting of Original Communications, Specifications of Patent Inventions, and Selections of Useful Practical Papers from the Transactions of the Philosophical Societies of all Nations, etc* volume II, page 416
Murphy, W. S. 1937 *Father of the Submarine* William Kimber & Co.

Nash, J. R. 1976 *Darkest Hours* Nelson-Hall
Neal, W. 1992 *With Disastrous Consequences* Hisarlik Press
*New Annual Registers*
Niel C. 1880 *Lancet* **ii** 557–8

*Oxford Book of Local Verses* 1987 Oxford University Press

Partington, J. R. 1960 *A History of Greek Fire and Gunpowder* Heffer
Pearsall, R. 1976 *The Alchemists* Weidenfeld & Nicholson
Pearson, Hesketh 1934 *The Smith of Smiths* Hamish Hamilton
Plimpton, G. 1984 *Fireworks — A History and Celebration* Doubleday
Pudney, John 1974 *Brunel and his World* Thames and Hudson

Quennell, P. (editor) 1937 *Private Letters of Princess Lieven* John Murray

Raikes, Thomas 1856 *A Portion of the Journal* Longmans
Ramazzini, B. 1964 *Diseases of Workers* Hafner
Read, J. 1942 *Explosives* Pelican Books
Readers Digest 1975 *Strange Stories — Amazing Facts*
Reeve, H. (editor) 1988 *The Greville Memoirs* Longmans, Green & Co.
Rolt, L. T .C. 1966 *The Aeronauts — A History of Ballooning* Longman
Rolt, L. T. C. 1957 *Isambard Kingdom Brunel* Penguin Books
Rolt, L. T. C. 1970 *Victorian Engineering* Penguin Books
Rolt, L. T. C. 1982 *Red for Danger* David and Charles
Ross, J. E. (editor) *Radical Adventurer — The Diaries of Robert Morris* Adams and Dart

Sandestrom, G. E. 1963 *The History of Tunnelling* Barrie & Rockliff
Schafer, L. S. 1990 *Best of Gravestone Humor* Sterling Publishing Co. Ltd.
Scott, A. (editor) 1975 *Everyone a Witness* White Lion Publishing
Skinner, Rev. J. 1930 *Journal of a Somerset Rector* John Murray
Southey, R. 1951 *Letters from England* Crescent Press

Spry, E. and Huxham, J. *An account of a case of a man who died of the effect of the fire at Eddystone Light-house by melted lead running down his throat (Philosophical Transactions of the Royal Society of London* Volume XLIV page 477 *et seq* and page 483 *et seq)*
Stokes, F. G. (editor) 1931 *The Blecheley Diary of the Rev. William Cole* Constable

*The European Magazine* March 1807
*The Hazardous Cargo Bulletin*
Thomas, J. 1972 *The Tay Bridge Disaster — New Light on the 1879 Tragedy* David and Charles
Thomas, J. 1976 *The West Highland Railway* David and Charles

Unknown Editor 1905 *Pryings Among Private Papers* Longman, Green & Co.

Vickery, R. 1990 *The Use of Conkers during the First World War* (*Plant-Lore Notes and News* volume 11, pages 49–50)

Walker, J. H. 1889 *The Johnstown Horror or Valley of Death* S. M. Southard
Wallington, N. 1989 *Images of Fire — 150 Years of Firefighting* David and Charles
Welch, A. 1978 *Accidents Happen* J. Murray
Wilson, D. 1978 *The Tower* Hamilton
Williams, G. 1975 *The Age of Agony* Constable
Williams, L. P. 1965 *Michael Faraday — A Biography* Chapman and Hall
Wright, G. N. 1972 *Discovering Epitaphs* Shire Publications
Wyndham Lewis, D. B. 1930 *The Stuffed Owl* J. M. Dent and Sons

Zapp, J. A. 1962 *Industrial Hygiene and Toxicology: Retrospect and Prospect* ed F. A. Patty, Wiley Interscience

# INDEX

For Product Safety Concerns and Information please contact our EU representative GPSR@taylorandfrancis.com
Taylor & Francis Verlag GmbH, Kaufingerstraße 24, 80331 München, Germany

www.ingramcontent.com/pod-product-compliance
Ingram Content Group UK Ltd.
Pitfield, Milton Keynes, MK11 3LW, UK
UKHW021005180425
457613UK00019B/810

* 9 7 8 0 7 5 0 3 0 3 6 0 6 *